ROCKIN - BENCHMARKING THROUGH ROBOT COMPETITIONS

RoCKIn - Benchmarking Through Robot Competitions
http://dx.doi.org/10.5772/intechopen.69115

Contributors

Alessandro Saffiotti, Tijn van der Zant, Pedro U. Lima, Luca Iocchi, Gerhard K. Kraetzschmar, Daniele Nardi, Pedro Miraldo, Emanuele Bastianelli, Roberto Capobianco, Rainer Bischoff, Tim Friedrich, Sven Schneider, Nico Hochgeschwende, Giulio Fontana, Matteo Matteucci, Francesco Amigoni, Viola Schiaffonati, Andrea Bonarini

Published by InTech

Janeza Trdine 9, 51000 Rijeka, Croatia

Publishing Process Manager Viktorija Zgela
Technical Editor SPi Global
Cover InTech Design team

First published July, 2017
Printed in Croatia
Legal deposit, Croatia: National and University Library in Zagreb

Additional hard copies can be obtained from orders@intechopen.com

RoCKIn - Benchmarking Through Robot Competitions
p. cm.
Print ISBN 978-953-51-3373-5
Online ISBN 978-953-51-3374-2

For EU product safety concerns:
IN TECH d.o.o., Prolaz Marije Krucifikse Kozulić 3, 51000 Rijeka, Croatia,
info@intechopen.com or visit our website at intechopen.com.

Contents

The book authors and RoCKIn project members would like to thank:

* Anne Bajart, from the European Commission DG CONNECT, RoCKIn's Project Officer, for her continuous support to the project activities, and encouragement to go always one step further in fostering research and benchmarking through robot competitions in Europe;

* all the members of the RoCKIn Advisory and Expert Boards, listed in Chapter 1, whose advice and contributions in written documents and meetings were crucial to improve the project output quality.

Chapter 1

Foreword: The Impact of RoCKIn on Robotics

Alessandro Saffiotti and Tijn van der Zant

Additional information is available at the end of the chapter

http://dx.doi.org/10.5772/intechopen.70307

When the RoCKIn project was conceived, the RoCKIn consortium decided to have a pool of external experts who will observe the project and provide feedback and advice. When we were asked to be part of this pool, we accepted enthusiastically. The reason for our enthusiasm was simple: benchmarking robots is a difficult task, but we must learn how to tackle it if we want to advance the field. The goal might be clear, but the methods to achieve the goal are open for interpretation. Adapting an existing benchmark is cumbersome. RoCKIn aimed to provide a way to explore possibilities and to provide a new way of thinking about benchmarking robots. Because of this, our intuition was that RoCKIn was one of those few projects that can contribute to redefine the fabric of robotic research, in order to align it both with our increased scientific expectations and with the new demands from the industry. We wanted to follow this adventure closely.

Our intuition turned out to be right. Despite its short duration and its small financial footprint, RoCKIn has created a distinctive impact on its own community but also on the robotic research community at large, on technology transfer, and on the general public. Here is why.

1. Impact on the participants

Quite obviously, the first candidates to benefit from RoCKIn are those researchers who participated in the competitions organized by the project. Did they?

RoCKIn pushed teams to advance the state of the art in terms of robotic technology. It did so by setting a research agenda that included specific challenges and specific performance metrics. The performance of most teams at the final competition, in 2015, was very good, and the top teams were just impressive. The progress made since the 2014 competition was considerable. This shows that the bar has been put at about the right level: a bit beyond the state of the art, but not so high that real progress cannot be made from one year to the next. This is quite a remarkable feat by itself.

Another interesting place where we observed a remarkable improvement over the lifetime of RoCKIn was the communication between organizers and participants. At the 2014 competition, we had the impression that some teams perceived the organizers as a separate, almost antagonist, entity. After that, the RoCKIn consortium worked hard to establish a better communication policy, to make teams feel that they are part of one and the same joint effort as the organizers, e.g., by having a shared understanding of the goals of the event, making the teams aware of the organizational difficulties and involving them in some organizational decision. This strategy worked out very well. When we watched the 2015 event, the teams and the organizers gave us the impression to act as one, cohesive unit that had been working together for a long time. The importance of having an open and effective communication between organizers and teams is an important practical lesson coming from RoCKIn.

From a technical point of view, an area that was and remains critical is system integration. Even the top robots were relatively brittle, which may suggest that system integration was a bit ad hoc: this impression was confirmed talking with the teams. This lack of integration is, unfortunately, rather common at robot benchmarks where the focus is on getting the software modules to work at all. Often, the focus is on getting through the tests. In a later stage, generalization comes into focus. More experienced teams take this into consideration from the start. Specialized finite-state machine solutions were preferred to the use of general purpose task planners, which would have been more flexible but more complex. The start-up time of the robots was very long, suggesting that many things had to be started and connected manually. If RoCKIn would continue, a logical next step would be to encourage a more systematic and general approach to system integration, for instance, by having challenges that involve the run-time modification and restart of the system.

Beside the technical advances, a great impact of RoCKIn was in terms of training of young researchers. The competition rules indirectly pushed the teams to adopt the values held by the RoCKIn consortium: modularity of the software, flexibility of the system, and replicability of the experiments. When interviewing the teams in 2014, many noticed that they had not put enough emphasis on these aspects and regarded this as one of their main weak points to be corrected for the 2015 competition. We see this awareness as a positive educational achievement of RoCKIn: the above values are important both for the development of a science of robotic systems and for the transfer of robotic techniques to industrial applications.

2. Impact on the robotic research community

It is safe to claim that RoCKIn advanced the state of the art in terms of experimental methodology in robotic research. The work on benchmarking and evaluation is one of the strong scientific contributions of RoCKIn and probably the one that will give RoCKIn its strongest impact in the long term. The idea of the functional challenges is an important and innovative part of the RoCKIn competitions. In fact, our perception is that the RoCKIn competitions are meta-experiments aimed at testing different hypotheses about what can be a "meaningful" evaluation metric. A good example of this method is the matrix "function × tasks." The entries of this matrix were initially an a priori guess about the correlations between functionalities and tasks, but as data from the competitions became available, they were used to confirm or disconfirm those correlations. This is, in our opinion, a novel and very promising methodological approach to empirical evaluation of complex systems—whether they are robotic systems or not.

The RoCKIn competitions took inspiration from RoboCup, and there was an inherent risk for RoCKIn to be perceived as yet another RoboCup-like activity. We soon realized that RoCKIn has done a good job in avoiding this risk. The project has defined its objectives and its methodology clearly, and it has implemented this event in a way to put forward what we see as its three strong messages about what one can do through a competition: to systematically evaluate full robotic systems, to benchmark key robotic functionalities, and to foster scientific communication and cooperation.

Having the data from the competition runs, including ground truth, is a strong added value of RoCKIn. The teams we talked to were excited about having these data. In addition, RoCKIn has adopted a strong open policy, which we applaud: the collected log files and ground truth data are intended to be openly available to the entire scientific community, not only the RoCKIn teams, which will make a big difference in impact. The RoCKIn open policy includes the creation of fully instrumented test facilities accessible for use by the robotic community at large. It is our hope that these repository and test facility will live well after the end of RoCKIn and that the heritage of RoCKIn is properly taken over after the end of the project.

3. Impact on technology transfer

One of the stated goals of RoCKIn is to help technology transfer in advanced autonomous robotic systems. This is an ambitious goal as the gap to bridge is wide. The RoCKIn@Work competition is strongly shaped by this goal, but technical progress in that section has been slower than in RoCKIn@Home. In fact, the main contribution of RoCKIn to technology transfer has probably been to highlight some of the main technological barriers that make it difficult and that require substantial research investments to be overcome.

The main barrier can be summarized in one word: robustness. It was surprising to note that most teams did not pay much attention to execution monitoring. Almost invariantly, whenever a robot failed to grasp an object or placed it improperly, the error was neither noticed nor corrected by the robot, which continued execution until the entire task inevitably failed. Failure detection and repair are key to achieving robust execution in open environments, which is critical for marketability. RoCKIn has helped us to put it in the research agenda. The next step would probably be to extend the RoCKIn benchmarks to include long-term or repeated experiments that stress robust operation over extended periods of time in non-nominal conditions.

A telling example of how robustness should enter in the benchmarking equation is provided by WLAN. Rather unsurprisingly, there were glitches in the WLAN connectivity during the competitions, and the performance of several robots was affected dramatically by these glitches. This, in our opinion, should not happen. A dependable domestic or industrial robot should be able to cope with reduced WLAN connectivity while remaining safe and reasonably functional. The ability to deal with WLAN problems should be one of the aspects that is tested in a robotic competition (as it is done in the DARPA challenge) since this is essential to real autonomy and deployability.

4. Impact on the general public

Robotic competitions have a fundamental role to play in informing and educating the general public about the reality of robotic research, trying to correct the too many misconceptions about robots and robotics. A strong effort must be placed to ensure that the public outreach is extensive and carefully prepared. RoCKIn was only partially successful in this respect, and it has helped us understand that public engagement should be one of the top priorities for future competitions.

During the 2014 competition, the host organizations in Toulouse (LAAS and the *Cité de l'Espace*) put an exceptional effort on dissemination: many visitors attended the event, and a professional commentator did a great job in explaining what was on. Despite this, we feel that the public received an unsatisfactory view of robotic research. The audience often expected to see Hollywood types of action but was faced with research robots where often there was "little action to watch." This problem is pervasive throughout robotic benchmarking. The robots often still require careful dedication and are far from being multipurpose machines with general types of intelligence. Many tests target specific capabilities, which make it difficult to tell a story to the audience. There is no clear solution, but probably showing only the best capabilities of the robots, and providing the audience with more understanding of what's going on inside them, could increase the appeal of the benchmarks. Organizers could decide to only open the finals and not all the preliminary trials, or they could allow the teams to do dedicated public demos, designed to be informative to a general audience (this was done as a last-minute addition to the program). One might also consider adding rules or scoring points related to the entertaining value of robots or including a new task to "interact with the public." Finally, the venue should be designed to maximize excitement, stimulate curiosity, and make explanations readily available. Showing a visualization of the internal state ("mind") of the robot on a big screen might also improve public engagement, allowing visitors to understand what the robot is doing and why. It would provide a commentator plenty of opportunities to explain general interesting things about robotics. Teams might also find this type of monitoring useful: we forgot how many times we heard the sentence "I do not know why the robot is doing this!"

5. Recommendation for future competitions

One of the important heritages left by RoCKIn is a set of best practices, lessons learned, and recommendations for benchmarking in general and for future competition in particular. We hinted at some of them above, and many more are contained in the different chapters of this book. We end this Foreword with four general recommendations that came from our experience as "external observers."

The most important one is to *have more competitions*. RoCKIn demonstrated that there is room for different types of benchmarks. At the start of the RoCKIn experience, there was some skepticism about the usefulness of yet another benchmark in a field where others already existed. By the end of the project, it became clear that we are only at the beginning of understanding what it means to benchmark robots. We need more benchmark projects like RoCKIn.

The second recommendation is to *keep exploring the space of possibilities* for robotic benchmarking. It is not always possible, or practical, to adapt existing benchmarks, mostly due to the committed investments of the participating teams. Short spikes of exploration can give guidance to longer running benchmarks by showing the pros and cons of particular ideas. This is essential for the progress of related benchmarks.

The third recommendation is to *consolidate the best practices* of organizing benchmarks. This includes everything from the first brainstorms to the creation of the rules to the actual running of the benchmark. The dissemination and the measured impact of the dissemination are also of great interest for many researchers and others in the benchmarking community. The present book is a step in this direction.

The last recommendation is to *radically experiment with the audience*. To be effective, the public dimension should be taken into account at all stages: from deciding the schedule, to designing the venue, to setting the rules. It is difficult to make a benchmark with relatively dumb and slow robots interesting for a general audience, but it is not impossible. Human-robot interaction with the audience is an interesting research topic. Best practices with respect to entertaining the audience might provide a large boost to the public acceptance of robotic research.

6. In conclusion

Sometimes, we are faced with projects that do not make much noise, but have nonetheless a profound and durable impact on the way we work on robotics. RoCKIn is one of those projects. We have seen the start of a new way of investigation into benchmarking. RoCKIn has demonstrated that benchmarking is a valid research topic in itself and one of growing importance to research, development, and innovation in robotics. Benchmarks deliver the tools required to advance the field of robotics. RoCKIn delivered the tools to advance the field of robotic benchmarking.

Author details

Alessandro Saffiotti[1]* and Tijn van der Zant[2]

*Address all correspondence to: asaffio@aass.oru.se

1 AASS Cognitive Robotic Systems Lab, Örebro University, Örebro, Sweden

2 CEO at RoboLect, CTO at SIM-CI, the Netherlands

Chapter 2

The RoCKIn Project

Pedro U. Lima

Additional information is available at the end of the chapter

http://dx.doi.org/10.5772/intechopen.70011

Abstract

The goal of the project "Robot Competitions Kick Innovation in Cognitive Systems and Robotics" (RoCKIn), funded by the European Commission under its 7th Framework Program, has been to speed up the progress toward smarter robots through scientific competitions. Two challenges have been selected for the competitions due to their high relevance and impact on Europe's societal and industrial needs: domestic service robots (RoCKIn@Home) and innovative robot applications in industry (RoCKIn@Work). The RoCKIn project has taken an approach to boosting scientific robot competitions in Europe by (i) specifying and designing open domain test beds for competitions targeting the two challenges; (ii) developing methods for scoring and benchmarking that allow to assess both particular subsystems as well as the integrated system; and (iii) organizing camps to build up a community of new teams, interested to participate in robot competitions. A significant number of dissemination activities on the relevance of robot competitions were carried out to promote research and education in robotics, as to researchers and lay citizens. The lessons learned during RoCKIn paved the way for a step forward in the organization and research impact of robot competitions, contributing for Europe to become a world leader in robotics research, education, and technology transfer.

Keywords: robotics, robot competitions, benchmarking, domestic robots, industrial robots

1. Introduction

Robot competitions have proved to be an effective instrument to foster scientific research and push the state of the art in a given field [1–6]. Teams participating in a competition must identify best practice solutions covering a wide range of functionalities and integrate them into practical systems. These systems have to work in realistic settings, outside of the usual laboratory conditions. The competition experience helps to transfer the applied methods and tools to successful and high-impact real-world applications. By participating in robot competitions, young students are attracted to science and engineering disciplines. Through competition events, the relevance of robotics research is demonstrated to citizens.

However, some limitations have emerged in the past as well-established robot competitions matured:

- the effort required to enter the competition grows and may present a barrier for the participation of new teams;
- a gap between benchmarking complete systems in competitions and benchmarking subsystems in research may develop and limit the usefulness of the competition results to industry.

The goal of "Robot Competitions Kick Innovation in Cognitive Systems and Robotics" (RoCKIn) has been to speed up the progress toward smarter robots through scientific competitions. Two challenges have been selected for the competitions due to their high relevance and impact on Europe's societal and industrial needs:

- domestic service robots (RoCKIn@Home) and
- innovative robot applications in industry (RoCKIn@Work).

Both challenges have been inspired by activities and their corresponding leagues in the RoboCup community [6–8], but RoCKIn extended them by introducing new and prevailing research topics, such as interaction with humans, and networking mobile robots with sensors and actuators spread over the environment (remotely controlled lamps, IP camera, motorized blinds home automation devices in RoCKIn@Home; drilling machine, conveyor belt factory-mockup devices in RoCKIn@Work), in addition to specifying specific scoring and benchmark criteria and methods to assess progress.

The RoCKIn project addressed the competition limitations identified above by (i) specifying and designing open domain test beds for competitions targeting the two challenges and usable by researchers worldwide; (ii) developing methods for scoring and benchmarking through competitions that allow to assess both particular subsystems and the integrated system; and (iii) organizing camps whose main objective is to build up a community of new teams interested to participate in robot competitions (**Figure 1**).

All these aspects of the project and their main outcomes are summarized in this chapter. We start by referring in Section 2 the two competition events organized during the project lifetime (January 2013–December 2015), listing good practices for the organization of scientific competitions in general. Section 3 presents the three RoCKIn camps and explains how they reached their goal of building a community of teams involved in robot competitions. Dissemination of robotics and its positive societal impacts, from education to technology that helps humans, is a crucial activity often connected to robot competitions. Section 4 explains how RoCKIn handled dissemination. The major outcomes of the project are summarized in

http://dx.doi.org/10.5772/intechopen.70011

Figure 1. A view of the RoCKIn 2014 venue, showing the arenas and a press visit.

Section 5, and they include scoring methods and metrics to evaluate the performance of robot systems, benchmarks, test beds designed as a standard reference, and rulebooks following the best practices in scientific robot competitions. The chapter closes with Section 6 on the project impact and lessons learned, as assessed by the project team but also by its external advisors. This chapter provides a work plan for future research triggered by robot competitions, including novel results on benchmarking, formal languages to describe competition rules objectively, dashboards for visualization of the robot state, and robot competitions as a cradle for open innovation.

2. RoCKIn competitions

Competition events were the core of RoCKIn. The test beds were developed to serve as a reference design to be used in all competitions for a given challenge, while the camps were organized to prepare new and existing teams to achieve good performances during competitions.

Within the project lifetime, two competition events took place, each of them based on the two challenges and their respective test beds:

- RoCKIn Competition 2014, in *La Cité de L'Espace*, Toulouse, November 24–30, 2014: 10 teams (7 @Home, 3 @Work) and 79 participants from 6 countries.
- RoCKIn Competition 2015, in the Portugal Pavilion, Lisbon, Portugal, November 17–23, 2015: 12 teams (9 @Home, 3 @Work) and 93 participants from 10 countries.

Organizing each of the competition events followed and improved established **best practices** for the organization of scientific competitions, which are listed here for future reference:

(1) setting up a Technical Committee (TC) per challenge, mostly composed of senior researchers experienced with the competitions and the specific challenges, responsible for enforcing the application of the rulebook competition rules;

(2) setting up an Organizing Committee (OC) per challenge, composed of researchers familiar with all the technical requirements and implementation details of the specific challenges, responsible to prepare the infrastructure and the whole setup in ways compatible with the rulebook requirements, as well as to report on that to provide transfer of information to organizers of upcoming events;

(3) [TC + OC] issuing the call for participation, requiring teams to submit an application consisting of a four-pages paper (named as Team Description Paper) describing the team research approach to the challenge, as well as the hardware and software architectures of its robot system, and any evidence of performance (e.g., videos);

(4) [TC] selecting the qualified teams, from among the applicants, based on their scientific and technical merits and past competition performance;

(5) [TC] preparing/updating and delivering the final version of the rulebooks, scoring criteria, modules, and metrics for benchmarking about 4–5 months before the actual competition dates, after an open discussion period with past participants and the robotics community in general;

(6) [OC] building and setting up the competition infrastructure at the venue, including a vision-based motion capture system (MCS) for ground-truth data collection during benchmarking experiments, listing all data to be logged by the teams during the competitions for later benchmarking processing, and preparing USB pens to store the data of the actual runs of the team's robot system;

(7) [OC] preparing several devices and software modules required by the competition rules (e.g., referee boxes, home automation devices and their network, factory-mockup devices and their network, objects for perception and manipulation, visitor's uniforms and mail packages, audio files, and lexicon), and describing their characteristics and technical specifications in a wiki page where all teams can access information, including a list of frequently asked questions and their answers, to ensure consistent replies to similar questions;

(8) [OC] establishing a schedule for the competition days and their different components (including team set up days and repeated runs of task benchmarks and functionality benchmarks);

(9) [OC] preparing human referees to follow the teams, handle referee boxes, record scores, and all the other required tasks;

(10) [OC] preparing the communication materials (brochure, leaflet, roller banners, banners, t-shirts, merchandising, and schedule) for the media and general citizens and stakeholder (from academia and industry) visitors, and materials for teams (bags, badges, and schedule);

(11) [TC+OC] establishing the adequate number of teams awarded per competition category and preparing trophies for the competition awards;

(12) [OC] realizing the event, including the organization of visits from schools, and the availability of communicators who explain to the audience what is happening, using a simplified version of technically correct descriptions.

Best practices such as those listed above foster scientific progress (by regular rule revisions to push the challenge forward, based on feedback from participants and end-users of the challenge scenarios), enforce technically rich approaches by the teams (by selecting them based on a team description paper) and peer monitoring (by setting up a technical committee composed of participants and other experts in the field), and enable transferring information about the competitions set up to next events (through reports prepared by the OCs), while making sure that dissemination to the general public is highly valued.

The competition scoring system was deployed so as to favor performance consistency, by taking mean values of scores over several runs, rather than picking the best run. This is more adequate for benchmarking purposes, since each run of a team is designed to have the same conditions as the other team runs, and simultaneously awards the teams that can consistently achieve good performances over several runs. A possible drawback of this approach is that the ability of robot system to adapt to unexpected situations may not be fully tested. On the other hand, teams tend to improve their performance over time as they fix problems from previous runs in new runs.

3. RoCKIn camps

Robot competitions need participating teams. Setting up a team to participate in a competition requires technical knowledge about the challenges, but also teamwork skills and experience on working and solving problems under pressure, as well as on preparing the team participation well before the competition dates.

RoCKIn camps were planned to build a community of teams experienced with robot competitions, and in particular with the technical details of the RoCKIn rulebooks and test bed infrastructure (e.g., interfacing the networked devices, handling the referee boxes). Simultaneously, camps acted as 1-week school where European experts trained European students on advanced robotic topics relevant for the RoCKIn challenges, such as object recognition and manipulation, and speech understanding.

Three camps were organized as follows:

- RoCKIn Kick-off Camp, in Eindhoven, the Netherlands, June 28 to July 1, 2013, during RoboCup2013: 12 participants. The camp consisted of several lectures by the partners, on RoCKIn challenges and activities, covering subjects such as principles for benchmarking robotics, raising awareness and disseminating robotics research, and discussion on developing robotics through scientific competitions like RoboCup. In addition to the lectures, attendees got first-hand experience of demo challenges, tests, and hardware and software solutions during the RoboCup@Home and RoboCup@Work practical sessions.

- RoCKIn Camp 2014, in Rome, Italy, January 26–30, 2014: 19 teams (11 @Home, 8 @Work), corresponding to a total of 63 students and researchers from 13 countries. This camp was designed to support the preparation of (preferably new) teams to participate in RoCKIn@Home and RoCKIn@Work competitions, and featured guest lectures by Michael Zillich, Norman Hendrich, and Matthew Walter on vision-based pattern recognition, object and people detection, object grasping and manipulation, and Human-Robot Interaction in natural language.
- RoCKIn Field Exercise 2015, in Peccioli, Italy, at the ECHORD++ Robotics Innovation Facility, March 18–22, 2015: 42 participants divided in 9 teams (4 @Home, 5 @Work). The Field Exercise has been designed as a follow up of the previous RoCKIn Camp 2014, where most of the RoCKIn Competition 2014 best teams displayed their progresses and all participants improved their interaction with the RoCKIn scoring and benchmarking infrastructure.

The selection of camp participants allowed two kinds of applications: team and individual. Teams had to submit a technical report on their existing or proposed technical approach to the RoCKIn challenges, while individuals had to submit a personal *curriculum vitae*. Selected individuals were assigned to teams.

Though the purpose of the different camps was different—as can be understood from their summary above—they were all structured similarly, i.e., including lectures by experts in particular topics and hands-on experiments with the robots and the test beds. Mini-competitions and awards for the best teams were created, so as to encourage team commitment and performance during the hands-on sessions.

The 2015 Field Exercise was particularly interesting because it took place at the ECHORD++ Robotics Innovation Facility (RIF) of Peccioli, Italy, funded by the European Commission. Teams gained access to the state-of-the-art ECHORD++ domestic test bed and to the RoCKIn@Work industrial test bed, and had the chance to practice and improve their performance in the task and functionality benchmarks, thus showing the portability of the industrial test bed and the ability to set up different test beds all over Europe according to the RoCKIn rules. The domestic test bed was equipped with the RoCKIn ground truth system for data gathering and allowed teams to get detailed feedback on their performance. This way, public funding is leading to a network of RIFs, including RoCKIn test beds, existing ECHORD++ RIFs and other test beds recently certified by the RockEU2 project, within the frame of the new European Robotics League (ERL), where robotics researchers can go benchmark their newly developed algorithms.

4. Disseminating robotics

Robot competitions have a crucial role on disseminating robotics research to the academic and industrial stakeholder communities, attracting young people for science and technology careers, and showing to lay citizens the impact of robotics technology on societal developments. Thus, dissemination activities focusing on the relevance of robot competitions had an important role in RoCKIn. These activities can be organized in three major categories as follows:

- **Presence in the web and social media**: a web page regularly updated; Facebook page and Twitter account also regularly updated, especially during major project events, such as the camps and the competitions; videos summarizing the RoCKIn Camp 2014, the RoCKIn Field Exercise 2015, the RoCKIn Competitions 2014 and 2015 were produced and made available online on the RoCKIn website and RoCKIn YouTube channel; videos describing the main goals of the benchmarks involved in the RoCKIn challenges were also produced and made available on the RoCKIn website.
- **Publications, presence in major robotics conferences and workshop organization**:
    - one key paper about the scoring and benchmarking methods used and the project activities was published on the *IEEE Robotics & Automation Magazine* [9];
    - presence in several scientific conferences, exhibitions, and industrial fairs, such as RoboCup 2013 (Eindhoven), IEEE ICRA 2013 (Karlsruhe), IEEE/RSJ IROS 2013 (Tokyo), IEEE ICAR 2013 (Montevideo), ISR/ROBOTIK 2014 @AUTOMATICA 2014 (Munich), EuRoC Challenge Design Workshop (Munich, 2014), IEEE ICRA 2014 (Hong Kong), IEEE/RSJ IROS 2014 (Chicago), INNOROBO (Lyon), IEEE/RSJ IROS 2015 (Hamburg), and ICT 2015 (Lisbon). The latter won the award for the best booth in the TRANSFORM area.
    - workshops on robot competitions, co-organized with the euRathlon and the EuRoC projects, during the European Robotics Forums (ERFs) in Lyon (2013), Rovereto (2014), Vienna (2015), and Ljubljana (2016).
- **Event co-location**:
    - euRobotics AISBL decided to move, for the first time, the communication center of the European Robotics Week to *La Cité de L'Espace* and Toulouse during RoCKIn Competition 2014;
    - RoCKIn Competition 2014 satellite events: *Les Journées Nationales de la Robotique Interactive*—organized by LAAS/CNRS (French academic conference); Friendliness made in Midi-Pyrénées (academia/industry networking event); Robotics EU Regions: Tell Me Who You Are (workshop on EU Robotics clusters and regions); Meetings of euRobotics Technology Topic Groups;
    - RoCKIn Competition 2015 satellite events: ROBOT2015—2nd Iberian Robotics Conference; EU Robotics Clusters Workshop (for Portuguese companies)—leading later to the setup of the *Lisboa Robotics Cluster*.

As part of the technical dissemination outputs of the project, two test beds were designed and built according to the rulebook open-source specifications, being available for research visits by worldwide groups interested to benchmark their approaches:

- RoCKIn@Home test bed at the Institute for Systems and Robotics of Instituto Superior Técnico, U. Lisboa, Portugal.
- RoCKIn@Work test bed at Bonn-Rhein-Sieg University of Applied Sciences, Sankt Augustin, Germany.

Test bed details are provided in Chapters 2 and 3 of this book.

5. Major outcomes

An estimated number of approximately 100 participants took part in the different activities (Camps, Competitions) organized within RoCKIn's frame. Many of them were new to robot competitions. Thus, one of the RoCKIn's top contributions was to build a larger community of robotics researchers interested in competitions in Europe.

Novel scientific and technological results are among RoCKIn's major outcomes:

- **Scoring methods and metrics** to evaluate and compare performance of different robot systems designed to solve given challenges, both at the task and functionality levels.
- **Benchmarking methods and metrics** to study the impact of functionality performance on task performance.
- **Open source design specifications for the test beds and rulebooks** of each challenge, which take into consideration the scoring and benchmarking requirements, together with problems whose solution requires pushing the state of the art in robotics research.

Before we highlight the main contributions under the above three topics, a brief description of RoCKIn's benchmarking and scoring systems is in order (detailed later in Chapter 4 of this book).

RoCKIn's approach to benchmarking experiments is based on the definition of two separate, but interconnected, types of benchmarks:

- **Functionality Benchmarks**, which evaluate the performance of hardware/software modules dedicated to single, specific functionalities in the context of experiments focused on such functionalities.
- **Task Benchmarks**, which assess the performance of integrated robot systems facing complex tasks that usually require the interaction of different functionalities.

Of the two types, Functionality Benchmarks are certainly the closest to a scientific experiment. This is due to their much more controlled setting and execution. On the other side, these specific aspects of Functionality Benchmarks limit their capability of capturing all the important aspects of the overall robot performance in a systemic way. More specifically, emerging system-level properties, such as the quality of integration between modules, cannot be assessed with Functional Benchmarks alone. For this reason, RoCKIn integrates them with Task Benchmarks.

In particular, evaluating only the performance of integrated system is interesting for the application, but it does neither allow to evaluate the single modules that are contributing to the global performance nor to point out the aspects needed to push their development forward. On the other side, the good performance of a module does not necessarily mean that it will perform well in the integrated system. For this reason, RoCKIn benchmarking targets both aspects and enables a deeper analysis of a robot system by combining system-level and module-level benchmarking.

System-level and module-level tests do not investigate the same properties of a robot. The module-level test has the benefit of focusing only on the specific functionality that a module is devoted to, removing interferences due to the performance of other modules, which are intrinsically connected at the system level. For instance, if the grasping performance of a mobile manipulator is tested by having it autonomously navigate to the grasping position, visually identify the item to be picked up, and finally grasp it, the effectiveness of the grasping functionality is affected by the actual position where the navigation module stopped the robot, and by the precision of the vision module in retrieving the pose and shape of the item. On the other side, if the grasping benchmark is executed by placing the robot in a predefined known position and by feeding it with precise information about the item to be picked up, the final result will be almost exclusively due to the performance of the grasping module itself. The first benchmark can be considered as a "system-level" benchmark, because it involves more than one functionality of the robot, and thus has limited worth as a benchmark of the grasping functionality. On the contrary, the latter test can assess the performance of the grasping module with minimal interference from other modules and a high repeatability: it can be classified as "module-level" benchmark.

5.1. Scoring methods and metrics to evaluate robot systems performance

The scoring framework for performance evaluation of robot systems in the Task Benchmarks of the RoCKIn@Home and RoCKIn@Work competitions is the same for all Task Benchmarks, and it is based on the concept of performance classes used for the ranking of robot performance in a specific task.

The performance class that a robot is assigned is determined by the number of achievements (or goals) that the robot reaches during its execution of the task. Within each class (i.e., a performance equivalence class), ranking is defined according to the number of penalties assigned to the robot. These are assigned to robots that, in the process of executing the assigned task, make one or more of the errors defined by a task-specific list associated with the Task Benchmark.

Performance classes and penalties for a Task Benchmark are task-specific, but they are grouped for all tasks according to three sets as follows:

- set of *disqualifying behaviors*, i.e., things that the robot must not do;
- set of *achievements* (also called goals), i.e., things that the robot should do;
- set of *penalizing behaviors*, i.e., things that the robot should not do.

One key property of this scoring system is that a robot that executes the required task completely will always be placed into a higher performance class than a robot that executes the task partially. In fact, penalties do not change the performance class assigned to a robot and only influence intra-class ranking.

It is not possible to define a single scoring framework for all Functionality Benchmarks, as for Task Benchmarks. These are specialized benchmarks, tightly focused on a single functionality, assessing how it operates and not (or not only) the final result of its operation. As a consequence, scoring mechanisms for Functionality Benchmarks cannot ignore how the functionality operates, and metrics are strictly connected to the features of the functionality. For this reason, different from what has been done for Task Benchmarks scoring methodologies and metrics are defined separately for each Functionality Benchmark of a competition.

In RoCKIn, Functionality Benchmarks are defined by four elements as follows:

- *Description*: a high level, general, description of the functionality.
- *Input/output*: the information available to the module implementing the functionality when executed, and the expected outcome.
- *Benchmarking data*: the data needed to perform the evaluation of the performance of the functional module.
- *Metrics*: algorithms to process benchmarking data in an objective way.

5.2. Benchmarking methods

The availability of both task and functionality rankings opens the way for the quantitative analysis of the importance of single functionalities in performing complex tasks. This is an innovative aspect triggered by the RoCKIn approach to competitions.

To state the importance of a functionality in performing a given task, RoCKIn borrows the concept of Shapley value from Game theory. Let us assume that a coalition of players (functionalities in the RoCKIn context) cooperates and obtains a certain overall gain from that cooperation (the Task Benchmark scoring in the RoCKIn context). Since some players may contribute more to the coalition than others or may possess different bargaining power (e.g., threatening to destroy the whole surplus), our goal is to calculate adequately how important is each functionality to the reach a given performance in a Task Benchmark.

5.3. Rulebooks, test beds, and datasets

The RoCKIn@Home test bed (see **Figure 2**) consists of the environment in which the competitions took place, including all the objects and artifacts in the environment, and the equipment brought into the environment for benchmarking purposes. An aspect that is comparatively new in robot competitions is that RoCKIn@Home is, to the best of our knowledge, the first open competition targeting an environment with ambient intelligence, i.e., the environment is equipped with networked electronic devices (lamps, motorized blinds, and IP cams) the robot can communicate and interact with, and which enables the robot to exert control over certain environment artifacts.

The RoCKIn@Home rulebook specifies in detail:

- The environment structure and properties (e.g., spatial arrangement, dimensions, and walls).

- Task-relevant objects in the environment, divide in three classes:
  - Navigation-relevant objects: objects that have extent in physical space and do (or may) intersect (in 3D) with the robot's navigation space, and which must be avoided by the robots.
  - Manipulation-relevant objects: objects that the robot may have manipulative interactions (e.g., touching, grasping, lifting, holding, pushing, and pulling).
  - Perception-relevant objects: objects that the robot must be able to perceive (in the sense of detecting the object by classifying it into a class, e.g., a *can*; recognizing the object as a particular instance of that class, e.g., a *7up can*; and localizing the object pose in a predetermined environment reference frame).

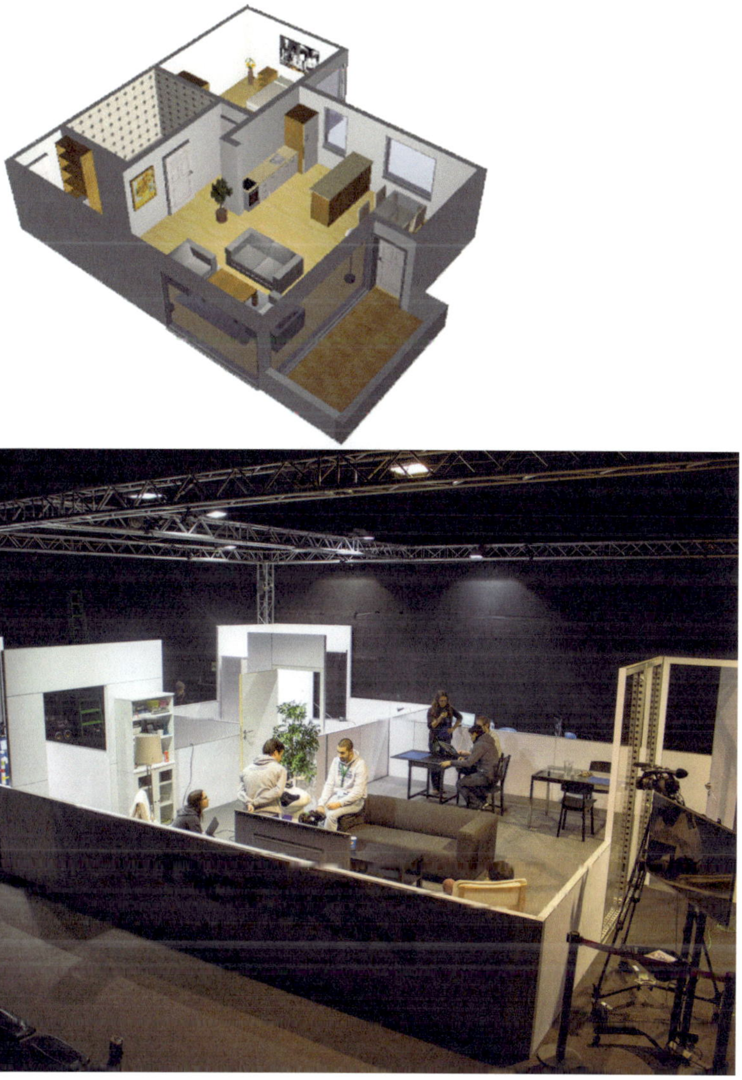

Figure 2. RoCKIn@Home test bed, including the trusses for the MCS cameras on the right.

During the benchmark runs executed in the test bed, a human referee enforces the rules. This referee must have a way to transmit his decisions to the robot, and receive some progress information, during the run and without interacting with the robot. To achieve this in a practical way, an assistant referee is seated at a computer and communicates verbally with the main referee. The assistant referee uses the main Referee and the Scoring and Benchmarking Box (RSBB). Besides basic starting and stopping functionality, the RSBB is also designed to receive scoring input and provide fine-grained benchmark control for functionality benchmarks that require so. In the future, it will be developed to provide also information to the public and the team about the evolution of the robot during the task.

The RoCKIn@Work test bed (see **Figure 3**) consists of the environment in which the competitions took place (the RoCKIn'N'RoLLIn medium-sized factory, specialized in production

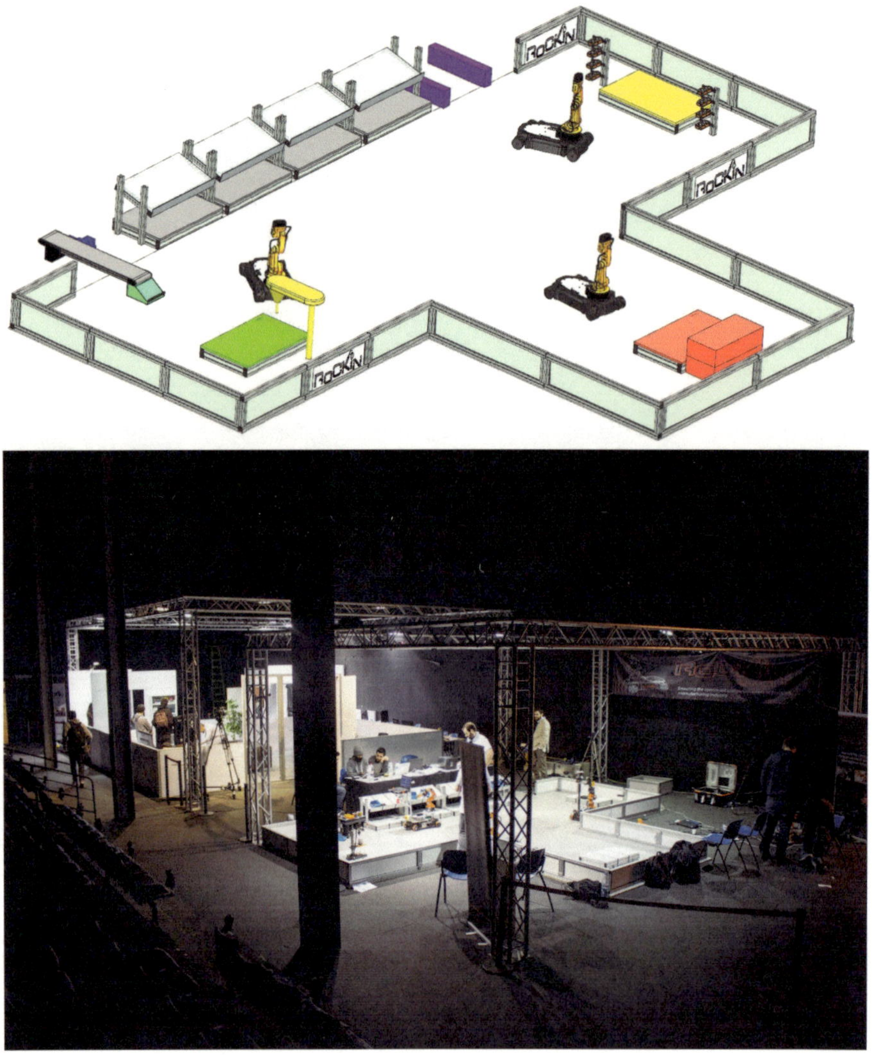

Figure 3. RoCKIn@Work test bed, including the trusses for the MCS cameras on the right.

of small- to medium-sized lots of mechanical parts and assembled mechatronic products, integrating incoming shipments of damaged or unwanted product and raw material in its production line), including all the objects and artifacts in the environment, and the equipment brought into the environment for benchmarking purposes. An aspect that is comparatively new in robot competitions is that RoCKIn@Work is, to the best of our knowledge, the first industry-oriented robot competition targeting an environment with ambient intelligence, i.e., the environment is equipped with networked electronic devices (e.g., a drilling machine, a conveyor belt, a force-fitting machine, and a quality control camera), the robot can communicate and interact with, and which allow the robot to exert control over certain environment artifacts like conveyor belts or machines.

The RoCKIn@Work rulebook specifies in detail:

- The environment structure and properties (e.g., spatial arrangement, dimensions, and walls).
- Typical factory objects in the environment to manipulate and to recognize.

The main idea of the RoCKIn@Work test bed software infrastructure is to have a central server-like hub (the RoCKIn@Work Central Factory Hub (CFH), equivalent to the RoCKIn@Home RSBB) that serves all the services that are needed for executing and scoring tasks and successfully realize the competition. This hub is derived from software systems well known in industrial business (e.g., SAP). It provides the robots with information regarding the specific tasks and tracks the production process as well as stock and logistics information of the RoCKIn'N'RoLLIn factory. It is a plug-in-driven software system. Each plug-in is responsible for a specific task, functionality, or other benchmarking module.

Both RoCKIn test beds include benchmarking equipment. RoCKIn benchmarking is based on the processing of data collected in two ways:

- internal benchmarking data, collected by the robot system under test;
- external benchmarking data, collected by the equipment embedded into the test bed.

External benchmarking data are generated by the RoCKIn test bed with a multitude of methods, depending on their nature. One of the types of external benchmarking data used by RoCKIn is pose data about robots and/or their constituent parts. To acquire these, RoCKIn uses a camera-based commercial motion capture system (MCS), composed of dedicated hardware and software. Benchmarking data have the form of a time series of poses of rigid elements of the robot (such as the base or the wrist). Once generated by the MCS system, pose data are acquired and logged by a customized external software system based on robot operating system (ROS).

Pose data are especially significant because it is used for multiple benchmarks. There are other types of external benchmarking data that RoCKIn acquires, however, these are usually collected using devices that are specific to the benchmark.

Finally, equipment to collect external benchmarking data includes any server which is part of the test bed and that the robot subjected to a benchmark has to access as part of the benchmark. Communication between servers and robot is performed via the test bed's own wireless network.

6. Impact and future expectations

RoCKIn's impact on the upcoming years is expected to be mostly supported by the scoring and benchmarking methods, as well as the test bed specifications, developed during the project lifetime, as they progressively (fully or partially) migrate to new European (European Robotics League[10]) and worldwide (RoboCup [6]) robot competitions. The research on benchmarking robot systems is also expected to be boosted by the introduced methods, as well as by exploiting RoCKIn major outcomes: the RoCKIn rulebooks, test beds, and datasets.

We have asked members of our Advisory Board on the potential impact of RoCKIn in several directions. The following are some of the important remarks, based on each advisor background, experience, and/or role in the scientific/industrial community:

- Prospects are bright for the preparation of a proposal for a European Robotics League, aimed at designing robot competitions as benchmarking experiments, where scoring methods encourage reproducibility and repeatability of experiments and provide methods to measure performance (e.g., error between actual outputs vs. ground-truth), while keeping the excitement of addressing a challenge and of competing with other teams to achieve the best solution to a common problem. Rules should foster developing functionalities that can be used and combined in the tasks. Algorithm and code-sharing repository of software modules per challenge should be developed.[1] Datasets from the competitions should be made available, while teams should be encouraged to log their runs and provide their datasets.
- RoCKIn@Work is very interesting to promote the evolution from a traditional industrial robotics to an emerging service robotics in manufacturing scenarios, where robots move, share the space and tasks with humans. This scenario has a huge potential in SMEs (thousands of companies all around Europe) but two aspects need to be addressed, the cost and the flexibility (e.g., regarding setup, easy programing methods, adaptability to changes at the product or at the production). The use of a perception system to detect objects including location is very interesting as a key functionality to achieve flexibility in manufacturing with robots. Perception is also a powerful tool to compensate the lack of accuracy of less specific robots and grippers.
- There is a fundamental gap between academic robotics research and robotics applications. A high technology readiness level can only be achieved for the latter if complete robotic system architectures are developed and evaluated in a strongly system-oriented manner. However, the involved efforts needed are often left out in academic research, although they are a prerequisite for transferring research results to real robotics applications. Robotic competitions can fill this gap, as they give reward to participants' efforts for systems development through good results. RoCKIn methodologies for robot systems integration from a scientific viewpoint enable more systematic benchmarking in competitions for intelligent robots and for transferring a rather "hands-on" way of organizing robot competitions to a more system-oriented research methodology. This pushes academic research toward methodologies for integration from a scientific standpoint.

[1]Teams are requested to make their code public on a voluntary basis, since there may exist situations, e.g., if they are using third party software or protected arts, where this may not be adequate.

- RoCKIn has also laid the basis for a yet not well-addressed aspect of future intelligent personal robots in industrial and home environments: standardization and certification. Meeting such standards with their robots, European robot manufacturers will be enabled to much better promote their high-tech robot developments in competition with other vendors on the international market.

The European Robotics League (ERL), whose foundations were laid out during discussions that took place during RoCKIn, started in early 2016 and aims to become a sustainable distributed format (i.e., not a single big event) which is similar to the format of the European Football Champions League, where the role of national leagues is played by existing *local* test beds (e.g., the RoCKIn test beds, but also the ECHORD++ RIFs), used as meeting points for "matches" where one or more teams visit the home team for a small tournament. This format exploits also arenas temporarily available during major competitions in Europe allowing the realization of larger events with more teams. According to this new format, teams could get "performance points" in a given challenge for each tournament they participate to, and they get ranked based on points accumulated over the year. Teams are encouraged to arrive 1–2 weeks before the actual competition/event so to participate in integration weeks where the hosting institution provides technical support on using the local infrastructure (referee boxes, data acquisition, and logging facility, etc.). *Local* tournaments take place in currently available test beds. *Major* tournaments are part of RoboCup and other similar events. The ERL has provided a certification process to assess any new candidate test beds as RIFs for both challenges, based on the rulebook specifications and the implementation of the proper benchmarking and scoring procedures. This will enable the creation of a network of European robotics test beds having the specific purpose of benchmarking domestic robots, innovative industrial robotics applications, and factory of the future scenarios.

The pool of ideas to extend and exploit RoCKIn scientific and technological results is large and exciting. We will list here the most relevant ones that came out from the 3 years of the RoCKIn experience, from the consortium members, the Experts Board members, and from Robotics researchers in general:

- The benchmarking infrastructure, both software and hardware, was an impressive and distinctive feature of RoCKIn with respect to other existing robot competitions and challenges. A standardized, and preferably low cost, hardware infrastructure with open source software for automated measurements and dataset dissemination, and guidelines for equipment and software set up, to be used as a reference for other competitions and research laboratories, is the way forward.

- Assess *robustness* in performance scoring—among other examples, the ability to deal with WLAN failures (or reduced bandwidth, or big latency) should be one of the aspects that is tested, since this is essential to real autonomy and deployability (namely in home scenarios), possibly penalizing excessive use of bandwidth.

- Advance toward the introduction of the semantic level, using semantic tags, i.e., all data should be accompanied by semantic meta data that described the intention of the robot actions, as well as the progress that the robot is making in this intention, at least according to what its own executor process assesses as progress, including the logging of the associated tolerances regarding the error of what the robot accomplishes with respect to the desired goal(s).

- In the scoring system, trace steps toward a better balance between human's judgment and satisfaction as the ultimate goal, and indicators that can be objectively measured, possibly including additional user-oriented metrics like acceptability, usability, or perceived utility.
- Develop (graphical) user interfaces and providing real-time data to fill the slots of a dashboard displaying information to the attending public, e.g., information about the state of the robot actions such as grasping an object and whether the robot thinks it actually has successfully grasped it—this will force the teams to monitor and diagnose the performance of their robot systems and not only producing and storing data.
- Use the RoCKIn approach as a playground for open innovation, where several teams contribute with components that need to be integrated in a "standardized" manner to build up a successful "mixed team"—domotics companies, Internet of Things research groups, care technology providers should be targeted and challenged to provide infrastructure and/or components.
- Start a community effort to develop a formal language to describe robotic scenarios, robotic tasks, and robotic benchmarks, with the goal of reducing the size and increasing the objectivity of the rulebooks, so as to describe domains and tasks in a compact but nonambiguous way.
- Provide the challenge rules with different levels of difficulty, so as to enable teams with different expertise levels to enter the competitions, e.g., encouraging undergraduate as well as PhD students and researchers from companies.
- Enforce the usage of computer vision and computer graphics (which are emerging and trending topics in industry) in some parts of the rules, e.g., favoring visual localization and mapping.
- Bring into play issues such as safety and privacy protection for robots working with aged people at home.

Appendix

Project information and contacts

The **RoCKIn consortium** is composed of the following partners:

- Instituto Superior Técnico, project coordinator
- Università di Roma "La Sapienza"
- Hochschule Bonn-Rhein-Sieg
- KUKA Roboter GmbH

- Politecnico di Milano
- InnoCentive

Advisory Board Members:

- Adam Jacoff, NIST, USA
- Bill Smart, Oregon State University, USA
- Bruno Siciliano, University of Naples Federico II, Italy
- Jon Agirre Ibarbia, Tecnalia, Spain
- Manuela Veloso, Carnegie-Mellon University, USA
- Oskar von Stryk, Technical University of Darmstadt, Germany
- XiaoPing Chen, University of Science and Technology of China, China

Experts board (reports on the competition events):

- Alessandro Saffiotti, Örebro University, Sweden
- Herman Bruyninckx, University of Leuven, Belgium
- Tijn van der Zant, University of Groningen, The Netherlands

RoCKIn contacts:

- Web: http://rockinrobotchallenge.eu
- E-mail: info@rockinrobotchallenge.eu
- Facebook: https://www.facebook.com/rockinrobotchallenge
- Twitter: @RoCKInChallenge

Author details

Pedro U. Lima

Address all correspondence to: pal@isr.tecnico.ulisboa.pt

Institute for Systems and Robotics, Instituto Superior Técnico, U. Lisboa, Portugal

References

[1] RoCKIn Project. Project Website [Internet]. 2014. Available from: http://rockinrobotchallenge.eu/ [Accessed: May 26, 2017]

[2] DARPA. DARPA Robotics Challenge [Internet]. 2015. Available from: http://archive.darpa.mil/roboticschallenge/ [Accessed: May 26, 2017]

[3] Behnke S. Robot competitions—Ideal benchmarks for robotics research. In: Proceeding IROS Workshop Benchmarks Robotics Research; 9-15 October 2006; Beijing, China. 2006

[4] Bonasso P, Dean T. A retrospective of the AAAI robot competitions. AI Magazine. 1997; **18**(1):11-23

[5] Bräunl T. Research relevance of mobile robot competitions. IEEE Robotics and Automation Magazine. 1999;**6**(4):32-37

[6] The RoboCup Federation. RoboCup [Internet]. 2016. Available from: http://www.robocup.prg [Accessed: May 26, 2017]

[7] Holz D, Iocchi L, van der Zant T. Benchmarking intelligent service robots through scientific competitions: The RoboCup@Home approach. In: Proceeding AAAI Spring Symposium. Designing Intelligent Robots: Reintegrating AI II; 25-27 March 2013; Palo Alto, California, USA. 2013. pp. 27-32

[8] Kraetzchmar G, Hochgeschwender N, Nowak W, Hegger F, Schneider S, Dwiputra R, Berghofer J, Bischoff R. RoboCup@Work: Competing for the factory of the future. In: Proceeding of RoboCup2014 Symposium; 25 July 2015; João Pessoa, Brazil. Springer; 2015

[9] Amigoni F, Bastianelli E, Berghofer J, Bonarini A, Fontana G, Hochgeschwender N, Iocchi L, Kraetzschmar G, Lima P, Matteucci M, Miraldo P, Nardi D, Schiaffonati V. Competitions for benchmarking: Task and functionality scoring complete performance assessment. IEEE Robotics & Automation Magazine. 2015;**22**(3):53-61. DOI: 10.1109/MRA.2015.2448871

[10] euRobotics. European Robotics League [Internet]. 2016. Available from: https://eu-robotics.net/robotics_league/ [Accessed: May 26, 2017]

Chapter 3

RoCKIn@Home: Domestic Robots Challenge

Luca Iocchi, Gerhard K. Kraetzschmar,
Daniele Nardi, Pedro U. Lima, Pedro Miraldo,
Emanuele Bastianelli and Roberto Capobianco

Additional information is available at the end of the chapter

http://dx.doi.org/10.5772/intechopen.70015

Abstract

Service robots performing complex tasks involving people in houses or public environments are becoming more and more common, and there is a huge interest from both the research and the industrial point of view. The RoCKIn@Home challenge has been designed to compare and evaluate different approaches and solutions to tasks related to the development of domestic and service robots. RoCKIn@Home competitions have been designed and executed according to the benchmarking methodology developed during the project and received very positive feedbacks from the participating teams. Tasks and functionality benchmarks are explained in detail.

Keywords: robot competitions, domestic robots, speech understanding, semantic mapping, person and object detection and recognition

1. RoCKIn@Home motivations and rules

With the goal of fostering scientific progress and innovation in cognitive systems and robotics, and to increase the public awareness of the current state-of-the-art of robotics in Europe, the RoCKIn project [1] developed RoCKIn@Home, a competition for domestic service robots. The competition was designed around challenges that are based on easy-to-communicate and convincing user stories, which catch the interest of both the general public and the scientific community. In particular, the latter aims at solving open scientific challenges and to thoroughly assess, compare and evaluate the developed approaches with competing ones.

The RoCKIn@Home competition hence aimed at bolstering research in service robotics for home applications, and to raise future capabilities of robot systems to meet societal challenges, like healthy ageing and longer independent living. To allow this to happen, competitions were designed to meet the requirements of benchmarking procedures and good experimental methods. The integration of benchmarking technology with the competition concept is one of the main goals of RoCKIn.

Behind the definition of the @Home benchmarks, we considered a scenario in which an elderly person, named 'Granny Annie', lives in an ordinary apartment. Granny Annie is suffering from typical problems of aging:

- She has mobility constraints and she gets tired fast;
- She needs to have some physical exercise;
- She needs to take her medicine regularly;
- She must drink enough;
- She must obey her diet;
- She needs to observe her blood pressure and blood sugar regularly;
- She needs to take care of her pets;
- She wants to have a vivid social life and welcome friends in her apartment occasionally, but regularly;
- Sometimes she has days not feeling so well and needs to stay in bed; and
- She still enjoys intellectual challenges and reads books, solves puzzles and socializes a lot with friends.

For all these activities, RoCKIn@Home is looking into ways to support Granny Annie in mastering her life. The context for performing such activities by technical systems is set in the subsequent scenario description.

The RoCKIn@Home scenario description is structured into three sections: environment, tasks and robots.

- The environment section specifies the environment in which tasks have to be performed. This information is also relevant for building testbeds and simulators.
- The tasks section provides details on the tasks the participating teams are expected to solve through the use of one or more robots and possibly additional equipment.
- The robot section specifies some constraints and requirements for participating robots, which mainly arise for practical reasons (size and weight limitations, for example) and/or due to the need to observe safety regulations.

2. The RoCKIn@Home environment

The goal of the RoCKIn@Home environment is to reflect an ordinary European apartment, with all its environmental aspects, like walls, windows, doors, blinds, etc., as well as common household items, furniture, decoration and so on. The apartment depicted in **Figure 1** serves as a guideline. More detailed specifications are given in the rule book. The following embedded devices are installed and are accessible within the apartment's WLAN:

- A networkable, camera-based intercom at the front door. It allows to see who is in front of the door;
- The lamps in the bedroom (e.g. on the bed stand) are accessible and controllable via network; and
- The shutters on the bedroom or living room window are accessible and controllable via network.

Figure 1. Model of the apartment used in the competitions.

3. Task benchmarks

Based on the user story described above, we defined three task and three functionality benchmarks. The latter represent basic functionalities that every robot should have, in order to successfully complete the tasks.

3.1. TBM1. Task benchmark *'Getting to know my home'*

The robot is told to learn about a new environment. It is supposed to generate a semantic map of the apartment within a limited time frame. How exactly to approach this task is left to the teams. For example, a team member may 'demonstrate' the apartment by guiding the robot through the apartment, pointing to objects and speaking aloud their names. Alternatively, a robot may explore the environment completely autonomously. The robot may also interrogate a team member about the names of objects or places. At the end of the environment learning phase, the robot must show through a behaviour the understanding of the environment.

The expected robot behaviour in this task is:

- *Phase 1: knowledge acquisition*. The robot in any way (through human-robot interaction (HRI) or autonomously or mixed) has to detect changes,[1] which may include: open or close doors connecting two rooms; moved pieces of furniture; or moved objects, possible objects are shown in **Figure 2**. In case of a HRI-based approach, a team member can guide the robot in the environment and show the changes with only natural interactions (speech and gesture). No input devices are allowed (e.g. touch screens, tablets, mouse, keyboard, etc.). At any time, teams can decide to move to Phase 2, even if not all the changes have been detected. However, the task in Phase 2 can refer only to objects acquired during Phase 1.
- *Phase 2: knowledge use*. The robot has to show the use of the new acquired knowledge. This phase is accomplished by executing a user command mentioning one of the items affected by the change. The user command must be given to the robot in a natural way. The preferred way is using speech interaction.

During *Phase 1*, the robot can move around in the environment for up to the maximum time limit of this task, possibly accompanied by the user (a team member) and interacting with him/her. The robot has to detect changes, and then it must represent them in an explicit format. In *Phase 2*, the robot is asked (e.g. by receiving a voice command) to move one of the changed objects recognized in *Phase 1* to a piece of furniture, also recognized in *Phase 1*. The accomplishment of the behaviour in *Phase 2* will be rewarded only if it refers to an object/piece of furniture that has been correctly reported in the output of *Phase 1*.

For scoring and ranking, we consider the following items. The set A of achievements for this task are:

- The robot detects the door with changed state;

[1]Before each task run, some random changes in the environment are made with respect to the nominal configuration given to the teams during the set-up days.

- The robot detects each piece of moved furniture;
- The robot detects each changed object; and
- The robot correctly executes the command given in *Phase 2*.

The set PB of penalized behaviours for this task are:

- The robot requires multiple repetitions of human gesture/speech;
- The robot bumps into the furniture;
- The robot stops working; and
- The robot was helped to manipulate an object.

Figure 2. Objects used in TBM1.

Additional penalized behaviours may be identified and added to this list if deemed necessary. The set DB of disqualifying behaviours for this task are:

- The robot hits Annie or another person in the environment; and
- The robot damages the testbed.

Additional disqualifying behaviours may be identified and added to this list if deemed necessary. These sets will be completed in later rule revisions.

3.2. TBM2. Task benchmark '*Welcoming visitors*'

This task assesses the robot's capability to interact effectively with humans and to demonstrate different behaviours when dealing with known and unknown people.

Granny Annie stays in bed because she is not feeling well. The robot will handle visitors, who arrive and ring the doorbell, as described in Chapter 4.

In all runs of this task, the four persons indicated above will ring the doorbell. The robot is thus required to deal with all the situations described above. However, the order in which the people will appear will be randomized for each run. Every visit will terminate before the next one. Pictures of Dr. Kimble are available, and images of the uniforms of both the Deli Man and the Postman are also given to the teams (see **Figure 3**).

The task involves handling several visitors arriving in any sequence, but separately from each other. The robot must be able to handle/interact with an outside camera. If a visitor has been admitted, the robot should guide him out after the visit.

The expected robot behaviour in this task is:

- *Phase 1: detection and recognition of the visitor*. Whenever a person rings the doorbell, the robot can use its own on-board audio system or the signal from the home automation devices to detect the bell ring(s). The robot has to understand who the person is asking for a visit, using the external camera. If the robot does not detect the ring call after three times, then

Figure 3. Visitor uniforms for TBM2.

the person will leave and the task will continue with the next person after a while. The robot can choose any way of opening the door, either using its manipulator or requesting a referee, a team member or the visitor to open the door (e.g. using speech).

- *Phase 2: greeting of the visitor.* For each detected visitor, the robot has to greet the visitor. In this spoken sentence, the robot has to demonstrate that it understood the category of the person.
- *Phase 3: executing the visitor-specific behaviour.* Depending on the visitor, the following behaviours are expected:
- Dr. Kimble: the robot allows the Doctor to enter and guides him/her to Annie's bedroom. Then, it waits until the Doctor exits the bedroom, follows him/her to the entrance door and allows the Doctor to exit;
- Deli Man: the robot allows the Deli Man to enter, guides the Deli Man to the kitchen, asking him/her to deliver the breakfast box on the table. Then, it guides the Deli Man back to the entrance door, and allows him/her to exit;
- Postman: the robot allows the Postman to enter, receives the postal mail (or ask the Postman to put it in the table in the hall) and allows him/her to exit; and
- Unknown person: do nothing.

After the execution of the visitor-specific behaviour, the robot should return to the initial position where it can receive the next visit.

For scoring and ranking, the seta of achievements, penalized and disqualifying behaviours for this task are those listed in Chapter 4.

3.3. TBM3. Task benchmark *'Catering for Granny Annie's comfort'*

This benchmark aims at assessing the robot's performance of executing requests about Granny Annie's comfort in the apartment.

The robot helps Granny Annie with her daily tasks throughout the day. After waking up in the morning, Granny Annie calls the attention of her service robot by touching a button on her tablet computer.[2] When the robot approaches her, Granny Annie uses spoken commands to ask the robot to operate on several home-automated devices, for instance, lifting the shutters, switching on a light, etc. Besides operating on home-automated devices, Granny can also ask the robot to further provide comfort, by looking for several of her belongings and bringing them back to her (see examples in **Figure 4**). There is no specific amount for the number of requests that Granny Annie has for the robot and the requests do not follow any specific order.

In the context of this task, a subtask is considered to be the resulting behaviour taken by the robot to accomplish something that Granny asked it to. In practical terms, if she asks the robot, for instance, to get her a cup, the resulting subtask is the process of looking for and

[2]An application for this purpose is provided by the @Home organization committee.

Figure 4. Objects used in TBM3.

bringing the cup back to her. In each run of this task, the robot will be asked to perform several subtasks. Granny Annie may only give one command at a time, and only after the robot executes the corresponding subtask another one may be given.

For each run of this task, in no specific order, the robot will be asked to operate the home devices and to find and bring back an object:

- Regarding the device operation, each team can choose whether the robot operates the devices with its manipulator, or over the home automation devices. The networked communication follows a pre-established common protocol which is specified by the organization committee; and
- A list of possible objects to be used is given to the teams, in advance. In addition, to ease the searching for objects, the likelihood of the position of the objects is also provided to the teams.

Afterwards, the robot will be given a finalizing command.

The expected robot behaviour in this task is:

- To reach the room where Granny Annie is located when she calls upon its service, approaching her in such a way that spoken communication is possible;
- The robot should then state its readiness to receive orders of subtasks to execute;
- When given a command, it should be confirmed in an appropriate way (e.g. by repeating it back to Granny Annie and asking if it was correctly understood). If the robot fails to understand a certain command after three tries, Granny Annie will move onto the next one; and
- The subtask corresponding to the given command should then be executed, and the robot should return to where Granny Annie is located.

This procedure should be repeated until Granny orders the robot to return to its idling position, concluding the task.

For scoring and ranking, we consider the following items. The set A of achievements for this task are:

- The robot enters the room where Granny Annie is waiting;
- The robot understands Annie's command(s);
- The robot operates correctly the right device(s);
- The robot finds the right object(s); and
- The robot brings to Annie the right object(s).

The set PB of penalized behaviours for this task are:

- The robot bumps into the furniture;
- The robot drops an object; and
- The robot stops working.

Additional penalized behaviours may be identified and added to this list if deemed necessary. The set DB of disqualifying behaviours for this task are:

- The robot hits Annie or another person in the environment;
- The robot damages or destroys the objects requested to manipulate; and
- The robot damages the testbed.

Additional disqualifying behaviours may be identified and added to this list if deemed necessary.

3.4. FBM1. Functionality benchmark '*Object perception*'

This functionality benchmark has the goal of assessing the capabilities of a robot in processing sensor data, in order to extract information about observed objects. All objects presented to the robot in this task benchmark are commonplace items that can be found in a domestic environment. Teams are provided with a list of individual objects (instances), subdivided in classes. The benchmark requires that the robot, when presented with objects from such list, detects their presence and estimates their class, instance and location. For example, when presented with a bottle of milk, the robot should detect a bottle (class) of milk (instance) and estimate its pose w.r.t. a known reference frame.

The set of individual objects, which will actually be presented to the robot during the execution of the functionality benchmark, is a subset of a larger set of available objects, here denoted as 'object instances' (examples of object instances, and their respective coordinates systems are shown in **Figure 5**). Object instances are subdivided into classes of objects that have one or more properties in common, here denoted as 'object classes'. Objects of the same

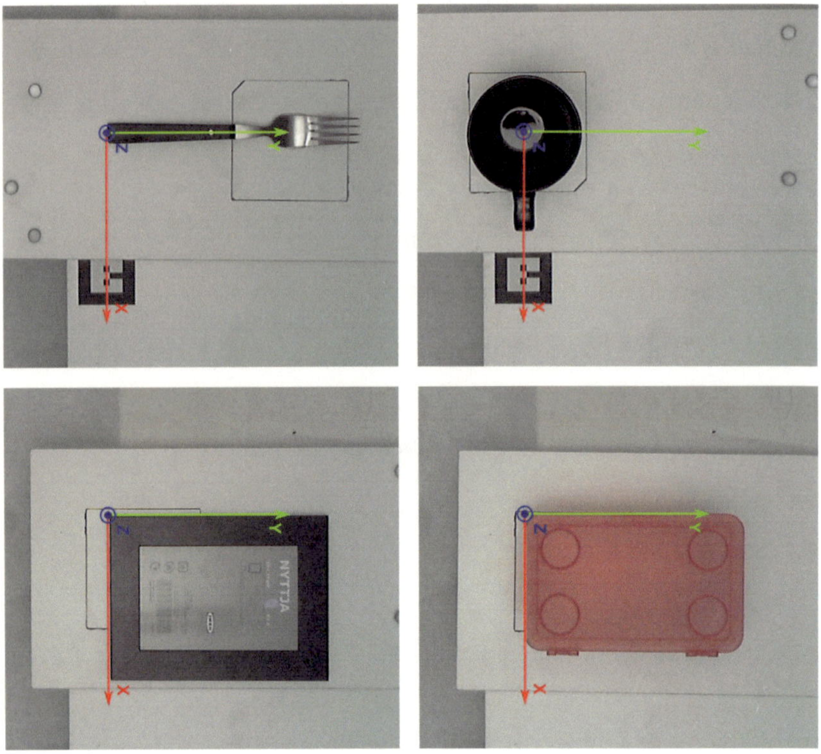

Figure 5. Object instances for FBM1.

class share one or more properties, not necessarily related to their geometry (for instance, a class may include objects that share their application domain). Each object instance and each object class is assigned a unique ID.

All object instances and classes are known to the team before the benchmark, but the team does not know which object instances will actually be presented to the robot during the benchmark. More precisely, the team will be provided with the following information:

- Descriptions of all the object instances;
- Subdivision of the object instances into object classes (for instance: boxes, mugs, cutlery); and
- Reference systems to each object instance (to be used to express object poses).

Regarding the expected robot behaviour, the objects it is required to perceive are positioned, one at the time, on a table (benchmark setup area) located directly in front of the robot. The actual pose of the objects presented to the robot is unknown before they are set on the table. For each presented object, the robot must perform:

- Object detection (i.e. class recognition): perception of the presence of an object on the table and the association between the perceived object and one of the object classes;

- Object recognition (i.e. instance recognition): association between the perceived object and one of the object instances belonging to the selected class; and
- Object localization (i.e. pose estimation): estimation of the 3D pose of the perceived object, with respect to the benchmark setup reference frame (given *a priori*).

These steps are repeated until the time runs out, or the maximum number of objects has been processed.

The evaluation of the performance of a robot according to this functionality benchmark is based on:

1. The number and percentage of correctly classified objects;
2. The number and percentage of correctly identified objects;
3. Pose error for all correctly identified objects; and
4. Execution time (if less than the maximum allowed for the benchmark).

These criteria are in order of importance (since this functionality benchmark is primarily focused on object recognition). The first criterion is applied first and teams will be scored according to the common accuracy metrics. The ties are broken by using the second criterion, again applying accuracy metrics. Finally, if needed, the position error is evaluated as well.

3.5. FBM2. Functionality benchmark *'Navigation'*

This functionality benchmark aims at assessing the capabilities of a robot to autonomously navigate in a typical apartment, containing furniture and objects spread through the apartment's rooms. From a predefined starting position, the robot will receive a list of waypoints that it must visit, before reaching a goal position.

Teams will have to take into account the following changes between different runs:

- Distinct starting points, waypoints and goal positions;
- Different number of waypoints to reach the goal; and
- Different number of obstacles blocking the path.

Teams are required to set their robot on a specific starting position (given to the teams before each run). Then, the robot should behave as follows. It receives the start signal, as well as an ordered list of waypoints that it must reach. The robot must then follow the order in which the waypoints are sent, sending back a signal each time it reaches a waypoint. The evaluation of the navigation will take into account the following three items:

- The distance between the robot's position and the respective position of the waypoint. It will be accounted both the Euclidean distance between the waypoint and the robot, and the difference in the orientation;

- The time spent by the robot to go from each waypoint to the next waypoint; and
- The number of times that the robot hits each obstacle. If the robot hits the same obstacle more than once, it will count as multiple hits.

The functionality benchmark ends as soon as the robot reaches the last waypoint, the time available for the functionality benchmark expires or if the robot hard-hits an obstacle.

The objects that can be in the robot's path are divided as follows:

- *Static and previously mapped:* hardware already present in the house such as furniture, doors and walls. The teams should already have these obstacles mapped from set-up days. These items will not change during this functionality benchmark;
- *Static:* items Granny Annie left lying on the ground. The obstacles may be of different shapes and sizes, are not previously known by the teams and may be different in between runs; and
- *Dynamic:* Granny Annie's visitors. People moving inside the house. Obviously, the movement people will do is unpredictable.

Regarding the scoring and ranking, at each run and for each team, three metrics will be used to score the performance:

- Accuracy scoring will be based on the distance and the orientation errors. The mean of the distances between the robot and the target waypoint is computed and stored in A, while the difference in orientations computed and stored in B. After the computation of these accuracy scorings, they will be discretized and fitted in one of the following groups:
  - 1: $A < 10$ cm AND $B < 20°$;
  - 2: $A < 30$ cm AND $B < 45°$;
  - 3: $A < 50$ cm AND $B < 90°$; and
  - 4: $A < 80$ cm AND $B > 90°$;

A lower group number corresponds to the better performance. Therefore, teams will be ranked starting from group 1. Note that for a team to be placed in any of the groups, it must respect the limits for A and B. If a team has a score that does not fit any of the groups defined above (e.g. mean of the error above 80 cm), it will not receive scoring in the respective functionality benchmark run;

- If more than one team falls inside each of the previously defined group, the number of obstacle hits will be used as a tie breaker, where the team with less hits will be ranked first and so on. Note that hits will only be considered as a tie breaker, i.e. a team in group 2 will never be ranked before any team in group 1, despite of the number of hits; and
- If teams are still tied, time will be the decisive tie breaker.

3.6. FBM3. Functionality benchmark *'Speech understanding'*

This functionality benchmark aims at evaluating the ability of a robot to understand speech commands that a user gives in a home environment. A list of commands will be selected among the set of predefined recognizable commands (i.e. commands that the robot should be able to recognize within the tasks of the competition or in similar situations).

Each implemented system should be able to capture audio from an on-board microphone, to record the captured audio in a file and to interpret the corresponding utterance. A standard format for audio files will be chosen (e.g. WAV) and communicated to the teams in advance before the competition. The system should produce an output according to a final representation defined below. Such a representation will have to respect a command/arguments structure, where each argument is instantiated according to the command evoking the verb. It is referred to as Command Frame Representation (CFR) (e.g. 'go to the living room' will correspond to MOTION (goal:"living room")). Summarizing, for each interpreted command the following relevant information will be collected: an audio file, its correct transcription and the corresponding correct CFR.

Variations between different runs can be:

- Different complexity in the syntactic structures of the spoken commands;
- The use of complex grammatical features, as pronouns;
- The use of synonyms for referring to objects, and
- The use of sentences where more than one action is expressed, resulting in a composed command (e.g. 'take the bottle and bring it to me').

Furthermore, variation in the quality of the audio corresponding to the user utterances can be considered, as for representing more or less noisy conditions.

Some information about the lexicon (verbs and nouns of objects) used in the benchmark is made available to the teams before the competition. In order to evaluate the correct understanding of a command expressed in natural language (e.g. through a sentence), a semantic representation formalism based on semantic frames has been selected. Each frame corresponds to an action, namely, a robot command. A set of arguments is associated to each frame, specifying part of the command playing a particular role with respect to the action expressed by the frame. For example, in the command 'go to the dining room' the motion frame is expressed by the verb go, while the part of the sentence 'to the dining room' corresponds to the goal argument, indicating the destination of the motion action. The set of frames defined and selected for this benchmark are given to the times before the competition.

Composition of actions is also possible in the CFR, corresponding to more complex action as the 'pick and place' action, represented by a sequence of taking frame followed by a bringing frame (e.g. for the command 'take the box and bring it to the kitchen'). The grammar specifying the correct syntax for a CFR is also provided.

Regarding the expecting robot behaviour, it should be able to understand a command starting from the speech input. The robot should correctly transcribe the user utterance and recognize the action to perform, resulting in the correct command frame (e.g. MOTION for a motion command) and the arguments involved (e.g. the goal of a motion command). The output of the robot should provide the CFR format for each command. For each command uttered or for each audio file directly provided during the speech understanding functionality benchmark, the system should generate the corresponding transcription and the interpretation in the CFR format.

All the teams are evaluated on the same set of spoken sentences. These spoken sentences are divided in two groups: a first group is formed by pre-recorded audio files, and a second group by voice commands uttered by a user during the benchmark. The robots are disposed in a circle, and the audio are broadcast using a 360° speaker (or an equivalent structure of speakers) with high fidelity performance placed in the centre. In this way, all the robots receive the same audio at the same time.

All teams are required to perform this functionality benchmark according to the steps mentioned below:

1. Each team receives the audio files randomly selected among the predefined set. This subset is the same for each team in order to reproduce fair conditions in the evaluation. Only one button can be pressed (either a button in a graphical user interface (GUI) or a key in the keyboard) to start the benchmark;

2. For each audio file, the system should generate the corresponding interpretation in the CFR format, together with the correct transcription of the corresponding utterance. The time for this processing will be restricted to an amount that is communicated in advance by the organization committee; and

3. After a proper communication, a member of the organization committee pronounces some commands using a microphone. The audio is instantly reproduced using a loudspeaker, conveniently positioned to be equally distant from each robot involved in the benchmark. Each command will be given after an interval of about 15 s of silence from the previous one. During this second part of the test, a designated member of the team will be allowed to press a button of the robot PC once for each sentence uttered by the speaker.

After the test is completed, only one button can be pressed to stop the processing.

During the execution of the benchmark, the following data are collected:

- Sensor data (in the form of audio files) used by the robot to perform speech recognition;
- The set of all possible transcription for each user utterance;
- The final command produced during the natural language analysis process; and
- Intermediate information produced or used by the natural language understanding system during the analysis as, for example, syntactic information.

Regarding the scoring and ranking, different aspects of the speech understanding process are assessed:

- The word error rate on the transcription of the user utterances, in order to evaluate the performance of the speech recognition process.
- For the generated CFR, the performance of the system will be evaluated against the provided gold standard version of the CFR, which is conveniently paired with the analysed audio file and transcription. Two different performance metrics will be evaluated at this step. One measuring the ability of the system in recognizing the main action, called Action Classification (AcC), and one related to the classification of the action arguments, called Argument Classification (AgC). In both cases, the evaluations will be carried out in term of Precision, Recall and F-Measure. This process is inspired by the Semantic Role Labeling evaluation scheme proposed in [24]. For the AcC, this measures will be defined as follow:
  - Precision: the percentage of correctly tagged frames among all the frames tagged by the system;
  - Recall: the percentage of correctly tagged frames with respect to all the gold standard frames; and
  - F-Measure: the harmonic mean between Precision and Recall.

  Similarly, for the AgC, Precision, Recall and F-Measure will be evaluated, given an action *f*, as:
  - Precision: the percentage of correctly tagged arguments of *f* with respect to all the arguments tagged by the system for *f*.
  - Recall: the percentage of correctly tagged arguments of *f* with respect to all the gold standard arguments for *f*.
  - F-Measure: the harmonic mean between Precision and Recall.
  - Time utilized (if less than the maximum allowed for the benchmark).

The final score is evaluated considering both the AcC and the AgC. Only the F-Measure is considered for both measures, each one contributing for 50% of the score. The AgC F-Measure is evaluated for each argument, and the final F-Measure for the AgC is the sum of the single F-Measure of the single arguments divided by the number of arguments. This final score has to be considered as an equivalence class. If this score will be the same for two or more teams, the *WER* will be used as penalty to evaluate the final ranking. This means that a team belonging to an equivalence class cannot be ranked lower than one belonging to a lower one, even though the final score, considering the *WER* of the first is lower than the score of the second.

4. Robots and teams

The purpose of this section is twofold:

1. It specifies information about various robot features that can be derived from the environment and the targeted tasks. These features are to be considered at least as desirable, if not

required, for a proper solution of the task. Nevertheless, we will try to leave the design space for solutions as large as possible and to avoid premature and unjustified constraints.

2. The robot features specified here should be supplied in detail for any robot participating in the competition. This is necessary in order to allow better assessment of competition and benchmark results later on.

4.1. General specifications and constraints on robots and teams

A competition entry may use a single robot or multiple robots acting as a team. At least one of the robots entered by a team must be mobile, and able to visit different task-relevant locations by autonomous navigation. Teleoperation (using touch screens, tablets, mouse, keyboard, etc.) of robots for navigation is not permitted (except when otherwise specified, e.g. in particular instances of task and functionality benchmarks). The robot mobility must work in the kind of environments specified for RoCKIn@Home, and on the kind of floors defined in the RoCKIn@Home environment specifications.

Any robot used by a team may use any kind of on-board sensor subsystems, provided that the sensor system is admitted for use in the general public, its operation is safe at all times and it does not interfere with other teams or the environment infrastructure. A team may use any kind of sensor system provided as part of the environment, by correctly using a wireless communication protocol specified for such purpose and provided as part of the scenario.

Any robot used by a team may internally use any kind of communication subsystem, provided that the communication system is admitted for use in the general public, its operation is safe at all times and it does not interfere with other teams or the environment infrastructure. A robot team must be able to use the communication system provided as part of the environment by correctly using a protocol specified for such purpose and provided as part of the scenario.

Any mobile device (especially robots) must be designed to be usable with an on-board power supply (e.g. a battery). The power supply should be sufficient to guarantee electrical autonomy for a duration exceeding the periods foreseen in the various benchmarks, before recharging of batteries is necessary. Charging of robot batteries must be done outside of the competition environment.

Any robot or device used by a team as part of their solution approach must be suitably equipped with computational devices (such as on-board PCs, microcontrollers or similar) with sufficient computational power to ensure safe autonomous operation. Robots and other devices may use external computational facilities, including Internet services and cloud computing to provide richer functionalities, but the safe operation of robots and devices may not depend on the availability of communication bandwidth and the status of external services.

All robots are checked by the organization committee for compliance with the specifications and constraints described in the rulebook. Teams will be asked to show the safety mechanisms of their robots and to demonstrate their use. A live demonstration is necessary: for example, pushing an emergency stop button while the robot is moving and verifying that the robot immediately stops. If the robot has other mechanical devices (e.g. a manipulator), their safety must be demonstrated as well. This inspection is done before the competition.

5. RoCKIn@Home research challenges and solutions

The development of the functionalities required by the tasks described in the previous section was very challenging for the teams, since it required not only realizing and testing robust solutions for each component, but also to properly integrate them in a fully working system. In this section, we briefly summarize the main research challenges that inspired the competition tasks and provide some comments about the adopted solutions.

For other features not described here, such as navigation and mapping, standard off-the-shelf components have been used by the teams.

5.1. Person and object detection and recognition

Person and object detection and recognition are important basic functionalities for service robots. In RoCKIn, TBM2 and FBM1 focussed on these topics.

TBM2 was designed to assess the ability of robots to properly understand the user with whom they are interacting and to provide the adequate behaviour according to the situation.

Many techniques are available in computer vision for face detection [2], face recognition [3], person modelling and people tracking [4]. However, their application on a robotic platform with limited on-board computation, real-time constraints and limited Internet connection for using cloud services makes this functionality very challenging.

During RoCKin competitions, person recognition was addressed in TBM2, where the robot was required to distinguish among four different kinds of people and to act accordingly. Images to be processed came from a fixed external camera (the same for all the teams) through a wireless link to the robot. Moreover, the robot can also decide to open the door and further examine the person with its on-board sensors.

This setup allowed teams to use some calibration procedure to identify the visitors according to some known features. For example, Dr. Kimble can be recognized through face recognition, while the Deli Man and the Postman by their uniforms.

Although this component may be considered quite straightforward and easy to implement, the integration in the entire system and some practical difficulties of the competition environment (e.g. acquiring images through a wireless channel in real time) required a very robust implementation.

Object recognition was specifically assessed in FBM1. Also, this test is significantly different from standard computer vision benchmark, since (1) the robot can move its sensors in order to reach a desirable viewpoint or integrate several views over time, and (2) position and orientation of recognized objects must also be estimated. Items to be recognized were available to teams during the set-up days before the test and, also in this case, the teams could benefit from calibration procedures. However, the test takes place in a physical environment (not through image dataset) and thus a variability introduced by different lighting conditions between calibration time and testing time must be considered and robustness to this variability is required to keep a high score.

5.2. Speech understanding

Speech understanding is also a fundamental feature of service and domestic robots, since spoken language is the most natural human-human communication means. Robots capable of understanding human language become accessible to a wider range of users, especially non-experts. This task is composed by two sub-tasks, namely, automatic speech recognition (ASR), that is the process of translating an audio signal into a written text, and (spoken) natural language understanding (NLU), that is the process of assigning a semantic interpretation to the transcribed text [5]. Many techniques are available to tackle ASR and NLU. For the first sub-task, it is possible either to rely on grammar-based method [6], or free-form methods [7]. For NLU, it is possible to rely on features embedded in the grammar framework [8, 9], or rely on data-driven methods [10, 11], where several machine learning techniques can be applied. Gold standards (i.e. ground truths) are necessary to evaluate the performances of both tasks. One of the most used metrics to evaluate ASR systems is word error rate [12], which measures the distance between a transcription hypothesis and the correct transcription. The NLU task instead is often evaluated using metrics derived from information retrieval, namely, Precision (P), Recall (R) and F-Measure (F1), over the semantic annotations.

FBM3 has been designed specifically to assess speech understanding capabilities of robotic platforms. In general, the task was to acquire a set of audio inputs of spoken commands, transcribe them and finally provide a semantic interpretation for each input, representing the actions and the related arguments of the intended command. Such interpretation had to be given according to a formalism inspired by frame semantics [13], specifically as it is represented in FrameNet [14]. Apart from this formalism, no further constraints have been given on the task, so that every team could develop its own system, either relying on a grammar-based method, or on data-driven ones. The benchmark was organized in two phases. In the first one, the audio input was presented to the team as audio files, bypassing the microphone acquisition. In the second phase, which was less controlled and more realistic, a live audio coming from a speaker needed to be acquired and analysed. Given the composite nature of the speech understanding task, it has been necessary to measure the performance of the two aforementioned sub-tasks to eventually evaluate the FBM3. WER has been used for ASR. Two factors have been instead measured for the understanding step: the action recognition (AcC), that is the ability of recognizing the sole actions (without arguments) expressed in a sentence, and the argument recognition (AgR), which takes into account also the action arguments. P, R and F1 have been evaluated for both AcC and AgC.

In order to provide a resource for designing, training and testing speech understanding systems, a corpus of spoken commands has been collected [11, 15]. Such resource has been incrementally build before and during the RoCKIn events (camps and competitions), through simulated or real interaction with robotic platforms. It is a collection of audio files of spoken commands gathered in diverse environmental conditions. Each command transcription is tagged with different levels of linguistic information, like morphology, part-of-speech tags and syntactic dependency trees. On top of that, semantic information is provided in terms of frame semantics. This semantic layer encodes the action intended in a command, together with its parameters. Although resources to evaluate either speech recognition [16] or natural language understanding [17, 18] for robotics have been developed in the past, this resource differs from them in many aspects. First of all, the provided linguistic information is made explicit and given according to linguistically supported theories (e.g. POS-tags, syntactic dependencies and semantic frames). Secondly, it covers all the linguistic processing steps, providing both audio files and annotations over the corresponding transcriptions. It can be thus used to train or design general linguistic modules of a natural language processing pipeline for robotics. Thirdly, it has been gathered in different phases, and thus it presents a high variability in terms of background noise, complexity of language structures and cardinality of the lexicon. These peculiarities were transferred inside the FBM3, making it definitely different from other benchmarks, specifically for the variability of the language, and the specificity of the adopted semantic formalism. Teams had to devise systems capable of dealing with complex syntactic structures, as well as unseen words. Moreover, the live acquisition phase put additional challenges in setting up suitable microphone configurations. Such difficulties led to poor performance during the first runs of the FBM3, which improved sensitively while going further in the competition, reaching final convincing performance from more than one team at the very end. Although some promising results have been achieved along the whole FBM3, there are still some aspects to explore, and issues to be tackled. An important feature of spoken interaction is dialogue. Robots should be able to deal with longer and more complex spoken interactions to appear more natural, being able, for example, to manage anaphora phenomena that may arise during longer interactions. Another crucial aspect is the acquisition of the audio. The audio can come, from several directions, according to the speaker positioning. Reaching a uniform performance on input coming from different points is for sure a challenge to address.

5.3. Semantic mapping

Semantic mapping is the incremental process of associating relevant information of the world (i.e. spatial information, temporal events, agents and actions) to a formal description supported by a reasoning engine [19], with the aim of learning to understand, collaborate and communicate. In particular, a semantic map is a representation that contains, in addition to spatial information about the environment, assignments of mapped features to entities of known classes [20]. Semantic maps should represent knowledge that can be used by a robot for reasoning and behaviour generation, thus enabling additional information to be inferred whenever the representation is associated with a reasoning or planning engine.

Multiple approaches have been proposed in the literature, characterized by an extreme heterogeneity of methodologies for representing learned maps—that prevents comparative evaluations, standard validation and evaluation procedures, and benchmarking strategies. For example, in Ref. [21] environmental knowledge is represented by anchoring sensor data to symbols of a conceptual hierarchy, based on description logic. The authors validate their approach by building their own domestic-like environment and testing the learned model through the execution of navigation commands. A multi-layered representation, ranging from sensor-based maps to a conceptual abstraction (an OWL-DL ontology), is generated in Ref. [22]. Except for individual modules, their experimental evaluation is mainly qualitative.

Instead, in Ref. [23], a conceptual map is represented as a probabilistic chain graph model, and Ref. [23] evaluate their method by comparing the belief of the robot of being in a certain location against the ground truth. In practice, none of the cited works can compare the performance of their semantic mapping method against each other.

For this reason, in Ref. [19] a formalization of a basic general structure for semantic maps is proposed, as the result of a generalization and intersection effort with respect to the representations adopted in the literature. This representation is proposed to play the role of a common interface among all the semantic maps, and can be easily extended or specialized as needed. Given two semantic maps of the same environment that implement this basic representation, it is at least possible to compare both the semantic and the geometrical parts of the representations [24]. In particular, given a ground truth, it is possible to define some error metrics that account for both the lack and inconsistency of stored information.

This evaluation approach has been applied in the scoring of the TBM1 test. More specifically, the teams have to provide at the end of the run a KB containing the semantic information about the environment acquired during the test. This KB is compared with a ground truth and the score is assigned by considering how many correct semantic labels are reported in the output KB. The use of this scoring methodology was extremely useful to compare different approaches of semantic mapping and, as mentioned above, can be further extended and used outside RoCKIn tasks.

Author details

Luca Iocchi[1]*, Gerhard K. Kraetzschmar[2], Daniele Nardi[1], Pedro U. Lima[3], Pedro Miraldo[3], Emanuele Bastianelli[1] and Roberto Capobianco[1]

*Address all correspondence to: iocchi@dis.uniroma1.it

1 DIAG, Sapienza University of Rome, Italy

2 Bonn-Rhein-Sieg University of Applied Sciences, Sankt Augustin, Germany

3 Institute for Systems and Robotics, Instituto Superior Técnico, Universidade de Lisboa, Portugal

References

[1] RoCKIn Project. Project Website [Internet]. 2014. Available from: http://rockinrobotchallenge.eu/ [Accessed: 26 May 2017]

[2] Zhang C, Zhang Z. A Survey of Recent Advances in Face Detection, Microsoft Research. Technical Report 2010-66. https://www.microsoft.com/en-us/research/wp-content/uploads/2016/02/facedetsurvey.pdf

[3] Bowyer KW, Chang K, Flynn P. A survey of approaches and challenges in 3D and multi-modal 3D + 2D face recognition. Computer Vision and Image Understanding. 2006;**101**(1):1-15. ISSN 1077-3142

[4] Yilmaz A, Javed O, Shah M. Object tracking: A survey. ACM Computing Surveys. 2006;**38**(4):13. ISSN 0360-0300

[5] de Mori R. Spoken language understanding: A survey. In: Furui S, Kawahara T, editors. IEEE Workshop on Automatic Speech Recognition & Understanding, ASRU 2007; December 9-13, 2007; Kyoto, Japan. IEEE; 2007. pp. 365-376

[6] Bos J. Compilation of unification grammars with compositional semantics to speech recognition packages. In Proceedings of the 19th International Conference on Computational Linguistics. Vol. 1. COLING '02; Stroudsburg, PA, USA; Association for Computational Linguistics; 2002. pp. 1-7

[7] Chelba C, Xu P, Pereira F, Richardson T. Large scale distributed acoustic modeling with back-off n-grams. IEEE Transactions on Audio, Speech, and Language Processing. 2013;**21**(6):1158-1169, IEEE Press

[8] Bos J, Oka T. A spoken language interface with a mobile robot. Artificial Life and Robotics. 2007;**11**(1):42-47

[9] Connell J. Extensible grounding of speech for robot instruction. In: Markowitz J, editor. Robots that Talk and Listen. De Gruyter, Germany; 2014

[10] Tellex S, Kollar T, Dickerson S, Walter MR, Banerjee AG, Teller S, Roy N. Approaching the symbol grounding problem with probabilistic graphical models. AI Magazine. 2011;**34**(4):64-76

[11] Bastianelli E, Castellucci G, Croce D, Iocchi L, Basili R, Nardi D. Huric: A human robot interaction corpus. In: Proceedings of the Ninth International Conference on Language Resources and Evaluation (LREC'14); Reykjavik, Iceland. European Language Resources Association (ELRA); 2014

[12] Popovic M, Ney H. Word error rates: Decomposition over pos classes and applications for error analysis. In: Proceedings of the Second Workshop on Statistical Machine Translation, StatMT '07; Stroudsburg, PA, USA. Association for Computational Linguistics; 2007

[13] Fillmore CJ. Frames and the semantics of understanding. Quaderni di Semantica. 1985; **6**(2):222-254

[14] Baker CF, Fillmore CJ, Lowe JB. The berkeley frameNet project. In: Proceedings of ACL and COLING. Association for Computational Linguistics; 1998. pp. 86-90

[15] Bastianelli E, Iocchi L, Nardi D, Castellucci G, Croce D, Basili R. RoboCup@Home spoken corpus: Using robotic competitions for gathering datasets. In: RoboCup 2014: Robot World Cup XVIII [papers from the 18th Annual RoboCup International Symposium; 15 July 2014; Joaõ Pessoa, Brazil. 2014c. pp. 19-30

[16] Bugmann G, Klein E, Lauria S, Kyriacou T. Corpus-based robotics: A route instruction example. In: Proceedings of Intelligent Autonomous Systems (IAS-8); 2004. pp. 96-103

[17] Dukes K. Train robots: A dataset for natural language human-robot spatial interaction through verbal commands. In: ICSR. Embodied Communication of Goals and Intentions Workshop; 2013

[18] MacMahon M, Stankiewicz B, Kuipers B. Walk the talk: Connecting language, knowledge, and action in route instructions. In: Proceedings of the 21st National Conference on Artificial Intelligence. Vol. 2. AAAI '06. AAAI Press; 2006. pp. 1475-1482

[19] Capobianco R, Serafin J, Dichtl J, Grisetti G, Iocchi L, Nardi D. A proposal for semantic map representation and evaluation. In: 2015 European Conference on Mobile Robots. IEEE; 2015. pp. 1-6. DOI: 10.1109/ ECMR.2015.7324198

[20] Nüchter A, Hertzberg J. Towards semantic maps for mobile robots. Robotics and Autonomous Systems. 2008;**56**(11):915-926. DOI: 10.1016/j.robot.2008.08.001

[21] Galindo C, Saffiotti A, Coradeschi S, Buschka P, Fernandez-Madrigal JA, Gonzalez J. Multi-hierarchical semantic maps for mobile robotics. In: 2005 IEEE/RSJ International Conference on Intelligent Robots and Systems; 2005. pp. 2278-2283. DOI: 10.1109/ IROS.2005.1545511

[22] Zender H, Mozos OM, Jensfelt P, Kruijff GJM, Burgard W. Conceptual spatial representations for indoor mobile robots. Robotics and Autonomous Systems. 2008;**56**(6):493-502. DOI: 10.1016/j.robot.2008.03.007

[23] Pronobis A, Jensfelt P. Large-scale semantic mapping and reasoning with heterogeneous modalities. In: 2012 IEEE International Conference on Robotics and Automation. IEEE; 2012. pp. 3515-3522. DOI: 10.1109/ICRA.2012.6224637

[24] Capobianco R. Interactive generation and learning of semantic-driven robot behaviours [Thesis]. Sapienza University of Rome

Chapter 4

RoCKIn@Work: Industrial Robot Challenge

Rainer Bischoff, Tim Friedrich,
Gerhard K. Kraetzschmar, Sven Schneider and
Nico Hochgeschwender

Additional information is available at the end of the chapter

http://dx.doi.org/10.5772/intechopen.70014

Abstract

RoCKIn@Work was focused on benchmarks in the domain of industrial robots. Both task and functionality benchmarks were derived from real world applications. All of them were part of a bigger user story painting the picture of a scaled down real world factory scenario. Elements used to build the testbed were chosen from common materials in modern manufacturing environments. Networked devices, machines controllable through a central software component, were also part of the testbed and introduced a dynamic component to the task benchmarks. Strict guidelines on data logging were imposed on participating teams to ensure gathered data could be automatically evaluated. This also had the positive effect that teams were made aware of the importance of data logging, not only during a competition but also during research as useful utility in their own laboratory. Tasks and functionality benchmarks are explained in detail, starting with their use case in industry, further detailing their execution and providing information on scoring and ranking mechanisms for the specific benchmark.

Keywords: robotics, robot competitions, benchmarking, domestic robots, industrial robots

1. Introduction

RoCKIn@Work is a competition that aims at bringing together the benefits of scientific benchmarking with the economic potential of innovative robot applications for industry, which call for robots capable of working interactively with humans and requiring reduced initial programming.

The following user story is the basis on which the RoCKIn@Work competition is built. RoCKIn@Work is set in the RoCKIn'N'RoLLIn factory—a medium-sized factory that is trying to optimize its production process to meet the increasing number of unique demands from its customers. RoCKIn'N'RoLLIn specializes in the production of small- to medium-sized lots of mechanical parts and assembled mechatronic products. Furthermore, the RoCKIn'N'RoLLIn production line integrates incoming shipments of damaged or unwanted products and raw materials. A key requirement to ensure the competitiveness of European industry is greater automation in a wide range of application domains which include flexible production processes that can easily be adapted to customer demands.

In RoCKIn@Work, robots will assist with the assembly of a drive axle—a key component of the robot itself and therefore a step towards self-replicating robots. Tasks include locating, transporting and assembling necessary parts, checking their quality and preparing them for other machines and workers. By combining the versatility of human workers and the accuracy, reliability and robustness of mobile robot assistants, the entire production process is able to be optimized.

RoCKIn@Work is looking to make these innovative and flexible manufacturing systems, such as that required by the RoCKIn'N'RoLLIn factory, a reality. This is the inspiration behind the challenge and the following scenario description.

Section 2 gives an oversight of the RoCKIn'N'RoLLIn factory and introduces all hardware and software elements that were used. Section 3 gives a detailed description of the task benchmarks, the way they have to be executed and the way they are scored. It further gives an explanation on the decisions that were taken to create the task benchmarks and how they differ from other benchmarks. Section 4 does the same for the functional benchmarks. The last section of this chapter gives a short summary and details some of the impressions from RoCKIn camps and competitions.

2. The RoCKIn@Work environment

This section introduces all hardware and software elements that are needed for the RoCKIn'N'RoLLIn factory to come to life. The description focuses on the elements themselves. A more detailed overview, especially on the software infrastructure, is given in Ref. [1].

2.1. The RoCKIn@Work testbed

The testbed for RoCKIn@Work, explained in detail in Ref. [2], consists of the environment in which the competition took place, including all the objects and artefacts in the environment, and the equipment brought into the environment for benchmarking purposes. An aspect that was comparatively new in robot competitions is that RoCKIn@Work is, to the best of our knowledge, the first industry-oriented robot competition targeting an environment with ambient intelligence, i.e. the environment is equipped with networked electronic devices the robot can communicate and interact with, which allow the robot to exert control on certain environment artefacts like conveyor belts or machines. **Figures 1a** and **b** show the testbed as it was used during the RoCKIn@Work competition 2015 in Lisbon.

http://dx.doi.org/10.5772/intechopen.70014

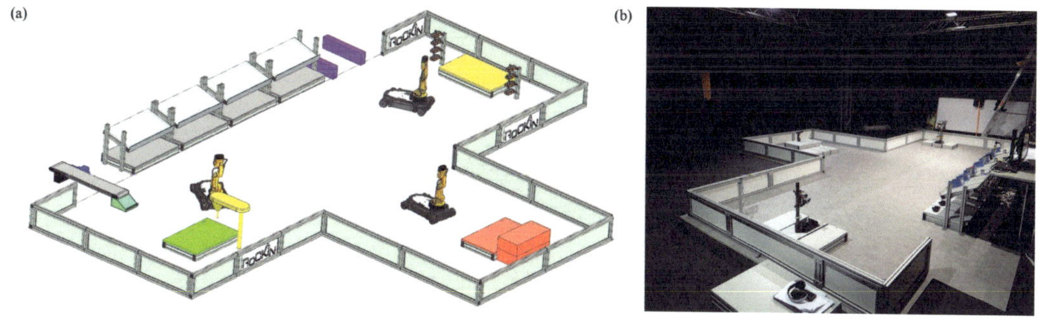

Figure 1. The RoCKIn@Work testbed. (a) Planned setup of the testbed and (b) Actual testbed at RoCKIn competition 2015.

2.2. Environment elements

To create an environment that closely resembles a real factory shop floor, a lot of different elements are necessary. In the case of RoCKIn@Work, the following elements are used:

- A set of shelves
- A force fitting workstation
- A drilling workstation
- A conveyor belt with a quality control camera mounted to it
- An assembly workstation
- A set of objects to be manipulated

Figure 2 shows an overview of the testbed elements.

Shelves are used to place objects. These objects range from a single object, e.g. a bearing, to containers storing multiple objects at once. The containers are so-called small load carriers. They are standardized containers in industry, originally meant to optimize the logistics chain of the automotive industry and their suppliers. The set of shelves area in RoCKIn@Work is a set of connected shelves and each shelf has two levels (upper level and lower level, see also **Figure 2**, lower right corner). The robot takes and/or delivers objects from the shelves (through the containers or directly onto shelves). The shelves are built from metal profiles and wooden panels, also something common on every factory floor. To make transportation and set-up easy, the construction of the shelves follows a modular design. Set-up and dismantling of all components can be done using a single Allen key. All components of the testbed fit on a single euro pallet after dismantling.

The force fitting workstation consists of a table for temporarily storing handled parts. The table itself is part of the force fitting machine, which can be operated by a robot or human worker. For this purpose, a drive was fixed to its structure. The drive is connected to a control board, which is attached to a Raspberry Pi microcontroller board running the necessary

Figure 2. Elements of the RoCKIn@Work testbed.

software to control the drive. On one side of the force fitting workstation, an assembly aid tray rack is placed. This assembly aid tray rack can be used to attach filled or unfilled aid trays, 3D-printed containers that can hold up to two bearing boxes, or finished assemblies. A more detailed description is given later in this section.

The drilling workstation consists of a storage area to store *file card* boxes and the drilling machine. The drilling machine is a simple model that can be purchased at a hardware store. Like for the force fitting machine, a drive with control board and a Raspberry Pi board were fixed to it so that the upwards/downwards motion can be controlled by a computer. Next to it, a conveyor belt is placed.

The conveyor belt transports parts from outside of the arena into the area. At the end of the conveyor belt, a quality control camera (QCC) was mounted. The camera is connected to the testbed's network and able to communicate with the robot through the Central Factory Hub (CFH; detailed below). Parts delivered into the arena fall down through guiders on an exit ramp in a predefined position where they can be taken by the robot.

The assembly workstation consists of a table, where a human worker can perform assembly of parts. The table features predefined areas where the robot can put boxes with supplies and pick up boxes with finished parts that had already been processed by the worker and need to be delivered elsewhere [3].

The objects present in the testbed can be subdivided into three classes as follows:

- Mechanical parts that have to be recognized and manipulated.
- Objects in the environment that have to be recognized and manipulated.
- Objects in the environment that have to be recognized only (because they are fixed to the environment, too heavy for the robot to lift or not relevant for the task).

Figure 3 shows the objects available in the testbed.

2.3. Central Factory Hub (CFH)

The main idea of the RoCKIn@Work testbed software infrastructure is to have a central server-like hub (the RoCKIn@Work Central Factory Hub) that serves all the services that are needed for executing and scoring tasks and successfully realizing the competition. This hub was derived from software systems, which are well known in industrial business (e.g. SAP). It provides the robots with information regarding the specific tasks and tracks the production process as well as stock and logistics information of the RoCKIn'N'RoLLIn factory. It uses a plug-in software architecture. Each plug-in is responsible for a specific task, for benchmarking or for other functionalities. A detailed description of the CFH and how it is utilized during RoCKIn and other robot competitions is given in Ref. [1].

2.4. Networked devices in the environment

The four networked devices described previously are used during execution of task benchmarks. This paragraph provides an overview on the capabilities of each networked device and its role in the related task. All networked devices can be operated through their connection to the Central Factory Hub. The software interface allows control either by the robot or through a graphical user interface by a human operator.

The force fitting machine is used for the insertion of a bearing into a bearing box. The force fitting process is performed by first inserting a bearing box with bearing on top of the bearing box. The placement process is executed with the help of an assembly aid tray. After the bearing box and bearing are properly placed, the force fitting machine is instructed to move down. Finally, the force fitting machine is instructed to move up again to make pick up of the processed item possible. The force fitting machine is used in the *Prepare Assembly Aid Tray for Force Fitting* task.

The drilling machine is used for drilling a cone sink in a cover plate. It is equipped with a customized fixture for the plates. Like for the force fitting machine, the drilling machine can be operated through its network interface with the CFH. The robot first has to insert the cover

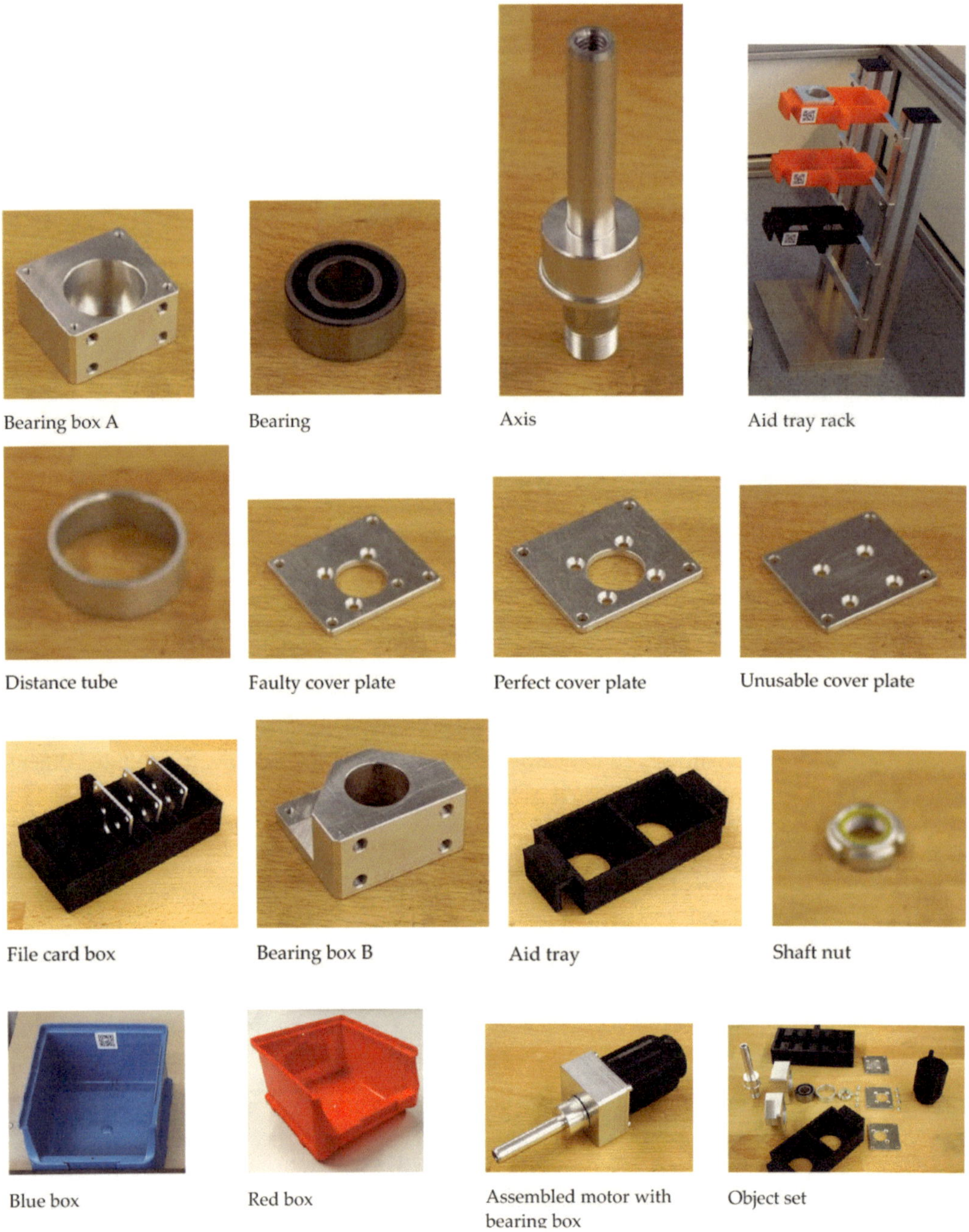

Figure 3. Objects in the RoCKIn@Work testbed.

plate into the fixture of the drilling machine. After that, the robot signals to the CFH to move the drill head down. Finally, the drill is moved up again and the drilled cover plate can be picked up. The drilling machine is used in the *Plate Drilling* task, specifically for the correction of a faulty cover plate.

The conveyor belt is used for delivering parts into the RoCKIn@Work testbed. At its end, it has a quality control camera to detect defects on the parts which are being delivered. The conveyor belt can be commanded, by the quality control camera, to move in both directions and to start/stop. It is not possible for the robot to directly interface with it. The conveyor belt is used in the *Plate Drilling* task.

The quality control camera or QCC is mounted above the conveyor belt and is used to acquire information about the quality of incoming cover plates delivered through the conveyor belt. The QCC also has the responsibility to deliver only a single cover plate through the conveyor belt (until the cover plate reaches the exit ramp of the conveyor belt) for each received command. After receiving a command, the QCC activates the conveyor belt until a cover plate is within the viewing range of the QCC. At this point, the QCC detects any defects of the cover plate. The conveyor belt keeps moving until it is being stopped by the QCC when the cover plate reaches the exit ramp of the conveyor belt. The QCC is used for the *Plate Drilling* task.

2.5. Benchmarking equipment in the environment

RoCKIn benchmarking is based on the processing of data collected in two ways as follows [3]:

- **Internal benchmarking data**: collected by the robot system under test.
- **External benchmarking data**: collected by the equipment embedded into the testbed.

External benchmarking data are generated by the RoCKIn testbed using a multitude of methods, depending on the nature of the data. One type of external benchmarking data used by RoCKIn is pose data about robots and/or their constituent parts. To acquire these, RoCKIn uses a camera-based commercial motion capture system (MCS) composed of dedicated hardware and software. Benchmarking data have the form of a time series of poses of rigid elements of the robot (such as the base or the wrist). Once generated by the MCS, pose data is acquired and logged by a customized external software system based on Robot Operating System (ROS). More precisely, logged data is saved as bagfiles created with the rosbag utility provided by ROS. Pose data is especially relevant because it is used for multiple benchmarks. There are other types of external benchmarking data that RoCKIn acquired. However, these are usually collected using devices specific to the benchmark. Finally, equipment to collect external benchmarking data includes any server which is part of the testbed and which the robot subjected to a benchmark has to access as part of the benchmark. Communication between servers and robot is performed via the testbeds' own wireless network. An extensive analysis on evaluation criteria and metrics for benchmarking is given in Ref. [4].

3. Task benchmarks

The concept of task benchmarks has already been introduced in Chapter 1. This section therefore describes details concerning rules, procedures, as well as scoring and benchmarking methods, which are common to all task benchmarks in the RoCKIn@Work competition

To make repeatability and reproducibility of the task benchmarks possible, teams have to follow a set of rules which are meant to lead to a more scientific benchmarking approach [5] instead of simply 'hacking' to get around a problem. To ensure a safe competition both for teams as well as the audience, every run of each of the task benchmarks has been preceded by a safety check. This is a very important aspect that often, especially with younger students, does not get sufficient attention. Much more often, a quick solution to a problem is found, but at the risk of injury. To avoid potential damage to the testbed or injury to participants, the team members must ensure and inform at least one of the organizing committee (OC) members present during the execution of the task that they have an emergency stop button on the robot which is fully functional. Any member of the OC can ask the team to stop their robot at any time, and such requests must be honoured immediately and swiftly. The OC member present during the execution of the task also makes sure that the robot is compliant with all safety-related rules and robot specifications defined in the rulebook. All teams are required to perform each task according to the steps mentioned in the 'Rules and Procedures' subsections for the tasks. During the competition, all teams are required to repeat a task benchmark multiple times. Each benchmark run is limited by a specified period of time.

During RoCKIn, benchmarking is of great importance. To gather as much information as possible and process the information later without error, guidelines on data storage had to be followed. This list presents the guidelines that are common to all task benchmarks. Specific information that has to be logged, but that only occurred during a single benchmark, is given later on in the description of the specific task benchmark.

- **Calibration parameters**: The calibration parameters for cameras have to be saved. This must also be done for other sensors that require calibration (e.g. Kinect), if a calibration procedure is applied instead of using the default values (e.g. those provided by OpenNI).
- **Notes on data storage**: The specific data that the robot has to save are described in the benchmark section. In general, some data streams (those with the highest bitrates) have to be logged only at time intervals when they are actually used by the robot to perform the activities required by the benchmark [3]. Thereby, system load and data bulk can be minimized. For instance, whenever a benchmark includes object recognition activities, video and point cloud data have to be logged by the robot only at times when performing object recognition.
- **Use of data**: The logged data is not used during the competition, in particular, it is not used for scoring. RoCKIn processed the data after the end of the competition. It was used for in-depth analysis and/or to produce datasets, published online, as given in ref. [3], for the benefit of the robotics community.
- **Where and when to store logged data**: Robots are required to store logged data in a specified format on an USB stick. The USB stick is given to the team immediately before the start of the benchmark by one of the RoCKIn partners and has to be returned (with the required data on it) at the end of the benchmark. All files produced by the robot that are associated with the execution of the benchmark have to be handed over.

Since benchmarking and data logging during robot competitions was a new concept, most teams were unaware of the implications this had on their system. To make sure that the data gathered lead to accurate results, teams were trained during the RoCKIn Camps and the RoCKIn Field Exercise on the principles of data logging. The camp and the field exercise usually took place early in the competition year, whereas the competition was held during the end of a year. The hands-on experience and help by the RoCKIn experts during the camp/field exercise led to most teams being able to correctly log the required data. Overall team performance was increased a lot and problems with the use of the software infrastructure provided by RoCKIn were minimized.

During the competitions, evaluation of the performance of a robot according to its task benchmark description is based on performance equivalence classes and they are related to whether the robot has performed the required task or not. The criterion determining the performance equivalence class of a robot is based on the concept of tasks requiring achievements, while the ranking of robots within each equivalence class is obtained by looking at the performance criteria. Specifically, the performance of a robot belonging to performance class N is considered better than the performance of a robot belonging to performance class M whenever M < N. In case, two robots fall into the same performance class, then a penalization criterion is used (penalties are defined according to task performance criteria) and the performance of the robot which received fewer penalizations is considered better. Finally, if two robots received the same number of penalizations, the performance of the robot which finished the task more quickly is considered better (unless not being able to reach a given achievement within a given time was already explicitly considered as a penalty). Thus, performance equivalence classes and in-class ranking of the robots are determined according to three sets as follows [3]:

- A set *A* of achievements, i.e. things that should happen (what the robot was expected to do).
- A set *PB* of penalized behaviours, i.e. robot behaviours that are penalized, if they happen, (e.g. bumping into the testbed).
- A set *DB* of disqualifying behaviours, i.e. robot behaviours that absolutely must not happen (e.g. hitting people).

Scoring was implemented with the following three-step sorting algorithm:

- If one or more of the elements of set *DB* occurred during task execution, the robot gets disqualified (i.e. assigned to the lowest possible performance class, called class 0), and no further scoring procedures are performed.
- Performance equivalence class X is assigned to the robot, when X corresponds to the number of achievements in set *A* that the robot accomplished.
- Whenever an element of set *PB* occurred, a penalty is assigned to the robot (without changing its performance class).

One key property of this scoring system is that a robot that executed the required task completely is always placed into a higher performance class than a robot that executed the task only partially. Moreover, penalties do not cause a change of class (also in the case of incomplete task execution).

3.1. Prepare assembly aid tray for force fitting

This task serves as an example for collecting and assembling parts from different locations. Additionally, teams can show their robot's capability in loading and unloading machines (a well-known industrial task). At the side of the aid tray—a container specifically built to hold two bearing boxes—unique identifiers are attached to uniquely identify the object. This task links to the concept of human robot collaboration (HRC), an idea that becomes more and more import in future factory environments. In this scenario, robots are not meant to take over human workers' jobs, but to support them and assist them with repetitive tasks or possibly unhealthy activities. This will be increasingly important in the future to react to increasing demand for customized products and to meet market demands.

3.1.1. Task description

The robot's task is to collect bearing boxes from stock (shelves) and to insert them into specialized aid trays. It is expected that the robot moves to the central station and registers with the Central Factory Hub. After receiving the task of *Assembly Aid Tray for Force Fitting*, the robot should locate the assembly aid tray in the shelf and proceed with identifying the identifiers on the assembly aid tray. The identifier encodes information like the assembly aid tray's serial number and the type of bearing box which can be fitted. Based on the examination of the assembly aid tray, the robot needs to find the correct bearing boxes in the shelves area. After finding the right bearing boxes, the robot has to record the identifiers of their containers, collect the bearing boxes and place them into the assembly aid tray. It can choose whether to deliver the bearing boxes collectively or individually based on its own reasoning. Once the assembly aid tray is filled with the bearing boxes, the trays can be loaded onto a force fitting machine, where the bearings are force fitted into the bearing boxes (see **Figure 4**).

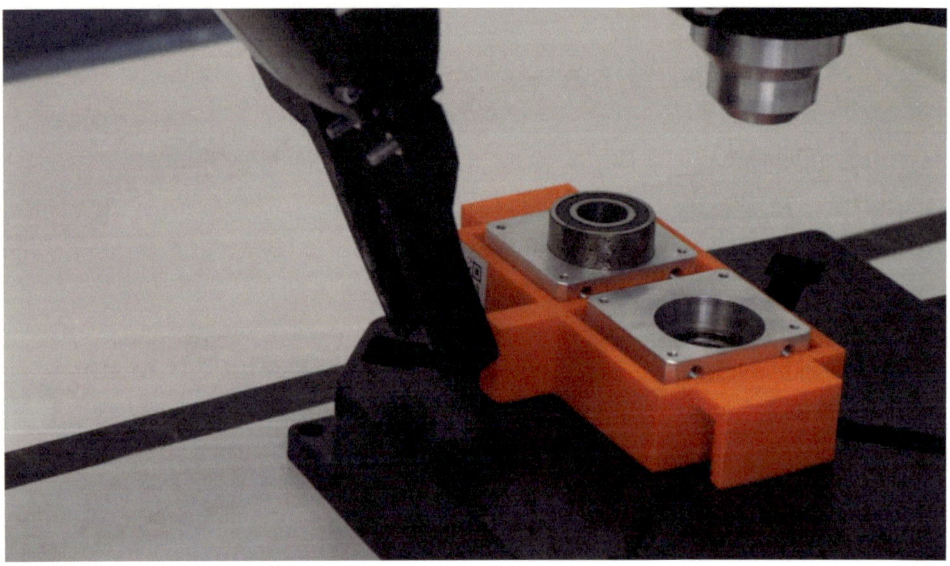

Figure 4. Force fitting of bearing into bearing box.

http://dx.doi.org/10.5772/intechopen.70014

When these steps of the process are completed, the robot gets a confirmation from the Central Factory Hub and has to perform a final examination of the finished product before its delivery. By scanning the identifiers as part of the task, the robot ensures continuous tracking of the production process and the parts involved. To make the challenge more realistic, some feature variation is possible. For example, the bearing boxes may come in different shapes. This variation is motivated by the modular concept of the final product, where the bearing box has to be inserted in different chassis. The robots are allowed to collect and insert the bearing boxes in the assembly aid tray individually or collectively.

3.1.2. Procedures and rules

Teams are provided with the following information:

- Description of the set of possible assembly aid trays and bearing boxes.
- Description and location(s) of the container(s) used for the bearing boxes.

During the task execution, the robot should perform the task autonomously and without any additional input. The task benchmark has to be carried out following these steps:

1. The robot is provided with multiple assembly aid trays and information about where the bearing boxes are stored.
2. Based on the identifier provided to the teams beforehand, the robot has to identify the appropriate bearing boxes it needs to put on the tray.
3. From the storage area, the robot has to pick the bearing boxes identified in Step 2 and insert them into the provided assembly aid tray.
4. The robot has to deliver the assembly aid tray to the force fitting workstation to be processed further.

Teams also have to be aware that an additional robot may be randomly moving in the arena which must be avoided by the participating robot. Although this randomizing element had been a possible feature variation, it was never actually applied in past competitions, because the dynamics of this feature could have had a negative impact on the repeatability and reproducibility of this benchmark.

3.1.3. Scoring and ranking

The performance equivalence classes used to evaluate the performance of a robot in this task benchmark are defined in relation to four different task elements. The first class is based on whether the robot correctly identified the assembly aid tray or not. The second class is defined by the robot correctly identifying the container or not. Class three uses the number of bearing boxes successfully inserted into the aid tray, and class four rewards the successful execution of the force fitting procedure. The fourth class encourages teams to try and solve the complete task instead of focusing on scoring only through pick-and-place actions.

The complete set A of achievements in this task included

- Correct identification of the assembly aid trays identifier
- Correct identification of the containers identifier
- Correct grasp of the assembly aid tray
- Correct grasp of the first bearing box
- Correct grasp of the second bearing box
- Correct insertion of the first bearing box into the aid tray
- Correct insertion of the second bearing box into the aid tray
- Correct delivery of the tray to the force fitting station
- Completely processing a part (from identifying to delivering)
- Cooperating with the CFH and networked devices throughout the task

At the end of the task benchmark, the team has to deliver the benchmarking data logged on a USB stick to one of the RoCKIn partners. If delivered appropriately and according to the guidelines, the team can score an additional achievement.

During the run, the robot should not bump into obstacles in the testbed, drop any object previously grasped or stop working. These behaviours are considered as behaviours that need to be penalized, and hence they are added to set *PB* of penalized behaviours.

The robot can demonstrate various behaviours that can lead to its disqualification. This includes, for instance, (a) if it damages or destroys the objects to be manipulated or the testbed; (b) if it shows extremely risky or dangerous behaviour; or (c) its emergency stop button is dysfunctional. Such disqualifying behaviours are added to the set *DB*.

3.2. Plate drilling

This task simulates handling of incomplete or faulty parts received from an external component supplier. The factory has to quickly react in such cases and create a process to correct the faulty parts. In principle, this task very closely corresponds to a real-world application. Not being able to manufacture components due to faulty incoming supplies can very quickly cost a lot of money. Especially, in times of 'just-in-time-manufacturing', where only small numbers of components are in stock, a faulty delivery can lead to a standstill of large parts of the production line. Being able to react fast and solve smaller issues yourself is crucial for manufacturing.

3.2.1. Task description

The cover plate of the bearing box has eight holes for connecting the motor with the bearing box. The four central holes need to have a cone sink. There are two possible defects of a cover plate which need to be accommodated in this task. The first case is where the supplier forgot to drill one of the cone sinks which results in a faulty cover plate. The faulty cover plates can be corrected by drilling the cone sink with the drilling machine available in the factory. The second case is where the cover plate is unusable and needs to be returned to the supplier for replacement. Examples of perfect, faulty and unusable cover plates are shown in **Figure 3**. The benchmark starts when the robot receives the task from the CFH. While performing the task, the robot has control over the QCC which allows it to regulate the flow of incoming cover plates. The robot has to send a command to the CFH to operate the QCC. Once the QCC receives a command from the CFH, the QCC activates the conveyor belt until a cover plate is placed on the exit ramp of the conveyor belt. During this process, the QCC detects the type of cover plate which is being delivered (either faulty or unusable) and sends this information to the CFH. Finally, the CFH broadcasts this information so that the robot knows that a faulty or an unusable cover plate was placed on the exit ramp of the conveyor belt. For each cover plate that arrives in the conveyor belt exit ramp, the robot needs to process them according to their fault status. An unusable cover plate needs to be delivered to the trash container box in the factory. For a faulty cover plate, the robot needs to perform correction by delivering it to the drilling machine (see **Figure 5**), operating the drilling machine to fix the missing cone sink, and placing the corrected plate in the file card box.

Figure 5. Retainer for faulty cover plate placement.

This benchmark also provides some possibilities for feature variation. For example, the sequence of faulty, unusable and perfect cover plates flowing over the conveyor belt is not fixed. This becomes relevant for the way a robot has to pick up cover plates from the exit ramp. If the robot needs to place the cover plate in the drilling machine for rework, specific positioning of its gripper may be required, while grip position is less important for the unusable plates, which end up in the trash container. The same holds for the orientation of the cover plate on the conveyor belt. In the first competition, it was planned that plate orientation is random, which would have led to more possibilities of grasping the plate from the exit ramp. For the second competition, this variation was not permitted in the spirit of repeatability. This is also true for the last variation. The number of plates delivered in each category (faulty, unusable and perfect) should have been randomized in each benchmark run, but it was decided that all teams should be able to execute exactly the same test. Furthermore, the solutions can vary depending on the sequence of activities being performed by the robot. The robot can choose to collect all cover plates from the conveyor belt first and process them collectively or perform the task for one cover plate at a time before collecting the next cover plate from the conveyor belt. The many variations possible through the robot's own reasoning and performance make the task benchmark challenging enough that none of the other variations were actually applied so far. The focus was instead set on repeatability, reproducibility and fairness between benchmark runs and teams.

3.2.2. Procedures and rules

Teams are provided with the following information:

- 3D CAD-textured models of the plates.
- Description of three different states of the plate (faulty, unusable and perfect).
- Location of objects related to the task.
- Commands for operating the QCC and the drilling machine.

During execution of the task, the robot should perform the task autonomously and without any additional input. The task benchmark is carried out by executing the following steps:

1. The robot controls the QCC until it receives feedback that a cover plate has arrived in the exit ramp of the conveyor belt.
2. The robot should pick up the cover plate and proceed with processing the cover plate as described in the next step.

According to the three possible fault types of the cover plate received, there are three possible (sequences of) actions to be executed by the robot as follows:

a. If the cover plate is perfect, the robot should place it in the file card box.

b. If the cover plate is unusable, the robot should drop it into the trash container box.

http://dx.doi.org/10.5772/intechopen.70014

c. If the cover plate is faulty, the robot should place the cover plate into the drilling machine, then perform the correction of the cover plate using the drilling machine, and finally place the corrected cover plate in the file card box.

In this benchmark, similar to the *Prepare Assembly Aid Tray for Force Fitting*, teams also have to be aware that an additional robot may be randomly moving in the arena. For the same reasons as mentioned in the preceding benchmark, this variation element was not yet applied during the benchmarking exercises and competitions.

3.2.3. Scoring and ranking

The performance equivalence classes used to evaluate the performance of a robot in this task benchmark are defined in dependence to three different categories. The first category is based on the number and percentage of correctly processed faulty cover plates. The second class refers to the number and percentage of correctly processed unusable cover plates. The third class uses only the execution time as measure (if less than the maximum time allowed for the benchmark was used by the robot). To encourage teams to try and solve the complete task in the *Plate Drilling* benchmark, it is also possible to score 'extra' achievements for the completion of a whole task (from request to delivery of a cover plate).

The complete set A of possible achievements in this task includes successful execution of

- Communication with the CFH throughout the test.
- Picking up a cover plate from the conveyor belt exit ramp.
- Placing an unusable cover plate in the trash container box.
- Complete handling an unusable cover plate (picking up an unusable cover plate from the conveyor belt exit ramp and placing it in the trash container box).
- Placing a faulty cover plate into the drilling machine.
- Performing the drilling of a faulty cover plate using the drilling machine.
- Complete handling a faulty cover plate (picking up a faulty cover plate from the conveyor belt exit ramp, placing it into the drilling machine, and performing the drilling of the faulty plate using the drilling machine).
- Picking up a corrected cover plate from the drilling machine.
- Placing a corrected cover plate into the file card box.
- Complete handling of a corrected cover plate (picking up a corrected cover plate from the drilling machine and placing it into the file card box).

At the end of the task benchmark, the team has to deliver the benchmarking data logged on a USB stick to one of the RoCKIn partners. If delivered appropriately and according to the guidelines, the team can score an additional achievement.

During the run, the robot should not bump into obstacles in the testbed, drop any object previously grasped or stop working. These behaviours are considered as behaviours that need to be penalized, and hence they are added to set *PB* of penalized behaviours.

The same disqualifying behaviours as in task benchmark *Plate Drilling* do apply for this task benchmark.

3.3. Fill a box with parts for manual assembly

This task benchmark reflects one of the primary requirements for a mobile robotic service assistant working together with humans. It is one of the most common tasks in industry: transporting parts from stock to the shop floor or to a human worker is very time consuming and requires well-planned logistics processes. For a human, it is cumbersome to check during his tour, if anything has changed or if he could pick up additional parts on his way. An automatic system has the advantage of direct communication to the shop floor management system and it can quickly respond and replan, if anything changes during production. The human worker can focus on his assembly task instead of worrying about parts arriving on time. This summarizes the idea behind this benchmark. The goal is to assist humans at a manual assembly workstation by delivering parts from different shelves to a common target location.

3.3.1. Task description

The robot has to fill boxes with parts for the final manual assembly of a drive axle. The task execution is triggered by the robot receiving a list of parts required for the assembly process from the CFH. It then proceeds by first collecting an empty box from the shelves, then collecting the requested parts (individually or collectively). When the parts have been placed in the box (see **Figure 6**), the robot delivers the box to the assembly workstation and provides the human worker with a list of parts in the box and a list of missing parts, if any. The boxes have no specific subdivisions; they may have foam material at the bottom to guarantee the safe transport. Thus, the robot has to plan the order of collecting the parts so that they can be easily arranged next to each other.

Feature variation in this task is kept to a minimum. Since it is a common task in industry, possible variations include different boxes, different parts and different locations for the parts. The planning and scheduling to best process the order is left to the teams. The benchmark itself aims for teams to show a good performance. All functional components of a robot system have to be used to solve the task, including navigation, object recognition, planning and manipulation. The benchmark still allows for more errors than the other benchmarks, e.g. position and orientation accuracy during navigation.

3.3.2. Procedures and rules

Teams are provided with the following information:

- The list of possible parts used in the task.

- Description of the box used for collecting the parts.
- Location of the parts in the arena.

During the execution of the task, the robot should perform the task autonomously and without any additional input. The task benchmark is carried out by executing the following steps:

1. The robot receives an order for a final assembly step of a product from the CFH containing a list of objects to be collected and delivered.
2. The robot plans the best path to the designated workstation, passing through each storage area where the required objects can be found.
3. The robot must move along the above path, collect the objects and deliver them to the designated area for final product assembly.
4. The Steps 2 and 3 above have to be performed for all the products in the list given in Step 1. Also, the robot has to follow, as much as possible, the priorities imposed by the philosophy of first-in/first-out when executing Steps 2 and 3.

Figure 6. Placing an assembly aid tray into small load container.

There may be multiple obstacles present in the scene that may block the direct path planned by the competing robot. If this is the case, the robot has to avoid all the obstacles or other robots during the execution of its task. To keep the benchmark repeatable and fair, new obstacles introduced to the testbed were positioned in the same place for all teams.

3.3.3. Scoring and ranking

The performance equivalence classes used to evaluate the performance of a robot in this task benchmark are defined in dependence to two different categories. The first class relates to the number of parts actually provided by the robot to the human worker, and the second class is based on how well the order of arrival corresponds to the desired one.

The complete set A of possible achievements in this task includes

- Communication with the CFH throughout the test.
- Picking up a required object (also the container) from its storage location.
- Placing the required objects into the container.
- Delivering a correctly filled container to the designated workstation.

At the end of the task benchmark, the team has to deliver the benchmarking data logged on a USB stick to one of the RoCKIn partners. If delivered appropriately and according to the guidelines, the team can score an additional achievement.

During the run, the robot should not bump into obstacles in the testbed, drop any object previously grasped or stop working. These behaviours are considered as behaviours that need to be penalized, and hence they are added to the set PB of penalized behaviours.

In this task benchmark the same disqualifying behaviours as in the previously mentioned task benchmarks are considered.

4. Functionality benchmarks

The concept of functionality benchmarks has already been introduced in Chapter 1. This section therefore describes details concerning rules, procedures, as well as scoring and benchmarking methods, which were common to all functional benchmarks in the RoCKIn@Work competition.

The basic execution and data logging guidelines as explained in Section 3 are also applied for functional benchmarks. Since communication with the CFH is of more importance in the functional than in the task benchmarks, teams need to follow additional rules. The easiest rule for the robot is to send a *BeaconSignal* message at least every second. This ensures that the CFH can detect whether a robot is still working or not. This also makes it possible to track when and how long a robot may have lost the connection to the CFH, for example, due to problems with the wireless network set-up. The second rule requests the robot to wait for a *BenchmarkState* message. It is supposed to start with testing the functionality as soon as the state received equals RUNNING. This allows the RoCKIn partners to set-up any elements necessary for the benchmark, without the possibility for a team to change anything during benchmark execution. The necessity to change elements during the run will be explained for each benchmark in the following sections. Other than in the task benchmarks, the third rule requires the robot to send the benchmarking data online to the CFH as soon as it is available. Specifically, the robot must send a message of type *BenchmarkFeedback* with the required data to the CFH. The robot should do this until the state variable of the *BenchmarkState* messages changes from RUNNING to STOPPED. The functionality benchmark ends when the state variable of the *BenchmarkState* message changes to FINISHED. The strong focus on online communication through the CFH guarantees a fair execution of the benchmark and less chance for error, e.g. as caused by human benchmark operators failing to switch parts in time.

4.1. Object perception

This functionality benchmark has the objective of assessing the capabilities of a robot in processing sensor data to extract information about observed objects. Objects presented to the robot in this functionality benchmark are chosen to be representative for the type of factory scenario that RoCKIn@Work is based on. Teams are provided with a list of individual objects (*object instances*), subdivided into object classes as described in Ref. [3]. The benchmark requires the robot, upon presentation of objects from such a list, to detect their presence and to estimate their class, identity and location. For example, when presented a segment of a T-section metal profile, the robot has to detect that it sees a profile (*class*), with a T-shaped section (*instance*) and its position with respect to the known benchmark set-up reference frame.

4.1.1. Functionality description

The objects that the robot is required to perceive are positioned, one at the time, on a table located directly in front of the robot. Depending on the set-up, this table can either be a separate table outside of the testbed or a workstation within the testbed. The poses of the objects presented to the robot are unknown until they are actually set on the table. For each object presented to the robot, it has to show performance in three distinctive areas as follows:

- Object detection
- Object recognition
- Object localization

The object detection part tests the robot's ability to perceive the presence of an object on the table and associate it to one of the object classes. Object recognition tests the ability to associate the perceived object with a particular object instance within the selected object class. Object localization tests the ability to estimate the 3D pose of the perceived object with respect to the surface of the table. **Figure 7** shows different objects mounted on small wooden plates

Figure 7. Objects used during the benchmark. The plate in the foreground is used to aquire the ground truth data.

which fit to the plate in the foreground in only one way. This allows to easily capture the ground truth data.

Feature variation for this functionality benchmark consists only of the variations given by the test itself: The variation space for object features is defined by the (known) set of objects the robot may be exposed to, and the variation space for object locations is defined by the surface of the benchmarking area where objects are to be located.

4.1.2. Procedures and rules

The concrete set of objects presented to the robot during the execution of the functionality benchmark is a subset of a larger set of available objects (*object instances*). Object instances are categorized into classes of objects that have one or more properties in common (*object classes*). Objects of the same class share one or more properties, not necessarily related to their geometry (for instance, a class may include objects that share their application domain). Each object instance and each object class are assigned a unique ID. All object instances and classes are known to the teams before the benchmark, but a team does not know which particular object instance will be presented to the robot during the benchmark. More precisely, a team is provided with the following information:

- Descriptions/models of all the object instances in the form of 3D textured models.
- Categorization of the object instances into object classes (for instance: profiles, screws and joints)
- Reference systems associated with the table surface and each object instance (for expressing object poses)

Object descriptions are expressed according to widely accepted representations and well in advance of competitions.

During the execution of the task, the robot should perform the task autonomously and without any additional input. The functionality benchmark is carried out by performing the following steps:

1. An object of unknown class and unknown instance is placed in front of the robot.
2. The robot determines the object's class, the instance within that class, and the 3D pose of the object, saving it in the required format.
3. The preceding steps are repeated until time runs out or 10 objects have been processed.

Since this test does not include a dynamic set-up and only a single functionality is tested, teams do not have to consider possible changes in the environment, e.g. a second robot presenting an obstacle for robot motion or changes of the lighting conditions.

4.1.3. Scoring and ranking

Evaluation of a robot's performance in this functionality benchmark is based on

- The number and percentage of correctly classified objects.
- The number and percentage of correctly identified objects.
- Pose errors for correctly identified objects as measured by the ground truth system.
- Execution time (if less than the maximum allowed for the benchmark).

As this functionality benchmark focuses on object recognition, the previous criteria are applied in order of importance. The first criterion is applied first and teams are scored according to their accuracy. Ties are broken by the second criterion, which still applies accuracy metrics. Finally, position error is evaluated as well. Since the position error is highly affected by the precision of the ground truth system, a set of *distance classes* is used. Remaining cases of ties are resolved by execution time.

4.2. Manipulation

This functionality benchmark assesses the robot's ability to grasp different objects. An object from a known set of objects is presented to the robot. After identifying the object, the robot needs to perform the grasping motion, lift the object and notify the CFH that it has grasped the object.

4.2.1. Functionality description

The robot is placed in front of the test area, a planar surface. A single object is placed in the test area and the robot has to identify the object and move its end effector on top of it. Then the

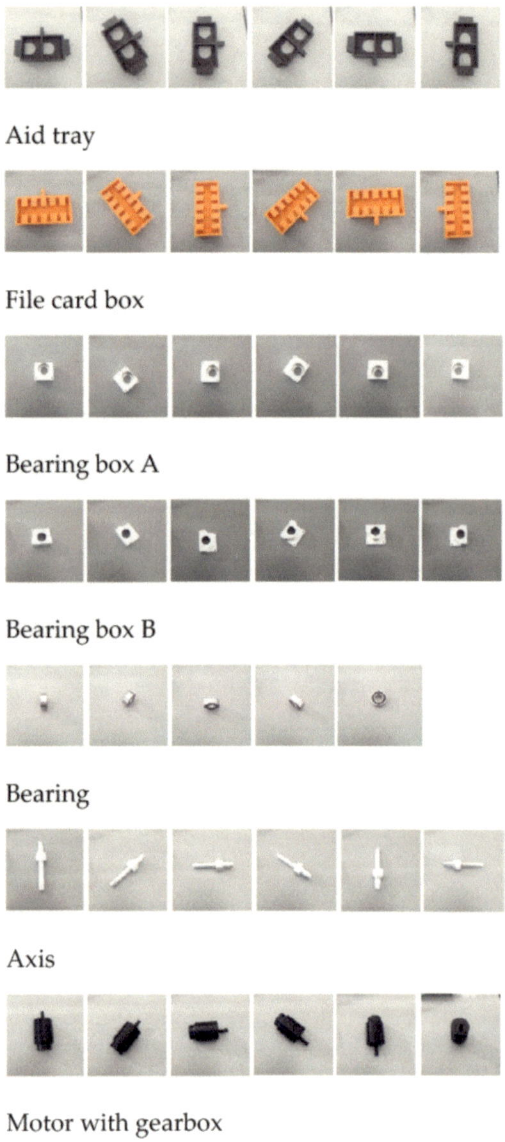

Figure 8. Object placement for the manipulation functionality benchmark.

robot should perform the grasping motion and notify that it has grasped the object. The task is repeated with different objects. So far, the following list of classes and instances of objects was used in the manipulation functionality benchmark:

- Containers
  - Assembly aid tray
  - File card box
  - Cover plates
- Bearing boxes
  - Bearing box type A
  - Bearing box type B
- Transmission parts
  - Bearing
  - Motor with gearbox
  - Axis

The objects used in the benchmark are selected from the complete set of parts used in the competition. The precise position of the objects differs in each test (examples are shown in **Figure 8**), which is necessary to avoid that grasping motions can be pre-planned by the teams and to ensure that the grasping motion really depends on the object presented. This test extends the object perception test by a manipulation part.

4.2.2. Procedures and rules

Teams are provided with the following information:

- The list of objects used in the functionality benchmark.
- Possible placements for each object used in the functionality benchmark.

During execution of the task, the robot should perform the task autonomously and without any additional input. The functionality benchmark is carried out by performing the following steps:

1. An object of unknown class and unknown instance is placed in the test area in front of the robot.
2. The robot determines the correct object class and object instance.
3. The robot grasps and lifts the object, then notifies the CFH that grasping has been performed.
4. The robot keeps the grip for a given time while the referee verifies the lifting.
5. The preceding steps are repeated with different objects.

For each object presented, the robot has to produce the result data consisting of the object's class name and instance name.

As this functionality benchmark does not foresee a dynamic set-up and only a single functionality is tested, teams do not have to consider possible changes in the environment, e.g. a second robot crossing its path or changes of the lighting conditions.

4.2.3. *Scoring and ranking*

Evaluation of a robot's performance in this functionality benchmark is based on

- The number and percentage of correctly identified objects.
- The number and percentage of correctly grasped objects; a grasp is considered successful when the object has no contact with the table any more.
- Execution time (if less than the maximum allowed for the benchmark).

Since this functionality benchmark focuses on manipulation, scoring of teams is based on the number of correctly grasped objects. A correct grasp is defined as the object being lifted from the table such that it is possible for the judge to pass his hand below it. For a grasp to be correct, the position has to be kept for at least 5 seconds from the time the judge has passed the hand below the object. The time the judge needs to verify the lifting of the object takes up to 10 seconds. In case of ties, the overall execution time is considered.

4.3. Control

This functionality benchmark assesses the robot's ability to control the manipulator motion and, if necessary, also the mobile platform motion, in a continuous path control problem. The ability to perform this functionality is essential in practice for precise object placement or for following a given trajectory in common industrial applications like welding or gluing. A path (or even a trajectory) is given to the robot. The robot has to follow this path with an end effector on its manipulator (examples shown in **Figure 9**).

The path is displayed on the table including a reference system. The external ground truth system measures the deviation of the path planned and executed by the robot from the given path by tracking a set of markers attached to the end effector.

4.3.1. *Functionality description*

The robot is placed in front of the test area, a planar surface. It first places its end effector on the top of a calibration point, then on the starting point with a fixed offset from the calibration point. At each of the two points, a manual calibration is performed by adjusting the position of the printed path (the table or sheet of paper). In order to synchronize the reference frames of the robot and the ground truth system, the robot detects the reference and starting point and then notifies the ground truth system about the positions of those points. The robot starts to follow the path and reports this to the CFH. After it finishes the movement, it has to signal this to the CFH.

Possible feature variations are the different paths the robot has to follow. In the second competition, where this benchmark was introduced first, the path was a simple line and sine. In future competitions, the path could be extended to become a general spline and it could be specified as trajectory including required velocity and acceleration vectors. The path is currently limited to the manipulator workspace, but can be extended well beyond this workspace in future competitions to force the mobile platform to move as well.

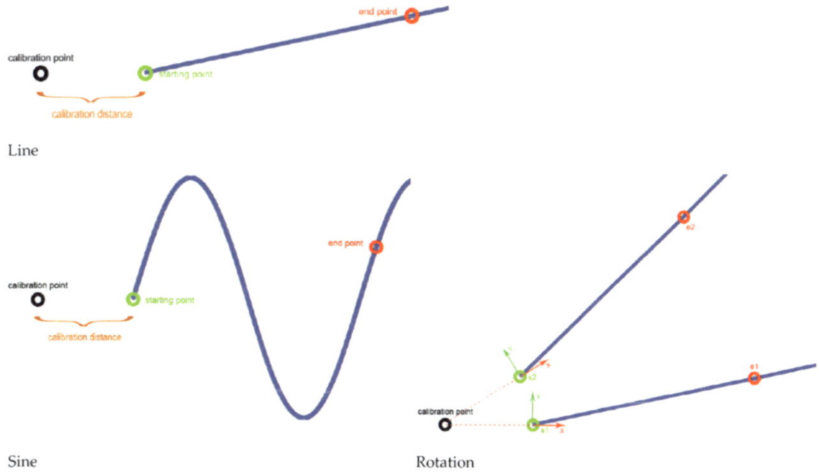

Figure 9. Example paths the robot had to follow.

4.3.2. Procedures and rules

Teams are provided with the following information:

- A mathematical description of a line in two-dimensional (2D) space.
- A mathematical description of a sine in 2D space.
- A list of generated points for both line and sine.
- Just before a competition run, the selection of the line or the sine for each run is published.

The path is provided including a starting point and a reference point next to it to enable calibration and synchronization with the ground truth system. Note that this task is not executed with a feedback from any vision sensor from the team, but only tests a pre-planned path and the online continuous path control ability of the robot!

During the execution of the task, the robot should perform the task autonomously and without any additional input. The functionality benchmark is carried out by executing the following steps:

1. The robot/team is provided with the selection of the specific path in advance.
2. The robot moves to the defined reference point within his coordinate system. Manual adjustment is then performed by a referee: the paper with the printed path is placed under

the actual position of the robot's end effector. (This is mainly important for the audience to get a visual feedback and to see the predefined path).

3. The robot moves to the defined starting point of the path, which is defined a few centimetres away with respect to the reference point. Another manual adjustment of the paper is then performed.
4. The CFH tells the robot when to start moving.
5. The robot moves its end effector along the path until the end point of the path is reached and reports the termination of path execution to the CFH.

4.3.3. Scoring and ranking

Evaluation of a robot's performance in this functionality benchmark is based on

- The overall deviation of the executed from the given path, measured in terms of the areas summing-up between the given and the executed path (constant deviations are eliminated).
- The number of completely executed path movements.
- Execution time (if less than the maximum allowed for the benchmark).

As this functionality benchmark focuses on control, the scoring of teams is based on the size of the area describing the deviation from given and executed path. In case of ties, the overall execution time is considered.

5. Summary

This chapter provides detailed information on the RoCKIn@Work competition. First, the competition, the concepts that build its foundation, and the intentions behind it are explained. After that, elements for building an open domain testbed for a robot competition set in the industrial domain are introduced and the most important aspects of benchmarking in competitions are outlined. The main part of this chapter covers in detail the three task benchmarks, *Prepare Assembly Aid Tray for Force Fitting, Plate Drilling* and *Fill a Box with Parts for Manual Assembly*, as well as the three functionality benchmarks, *Object Perception, Manipulation* and *Control.*

Author details

Rainer Bischoff[1], Tim Friedrich[1*], Gerhard K. Kraetzschmar[2], Sven Schneider[2] and Nico Hochgeschwender[2]

*Address all correspondence to: tim.friedrich@kuka.com

1 KUKA Roboter GmbH, Germany

2 Bonn-Rhein-Sieg University of Applied Sciences, Germany

References

[1] Niemueller T, Zug S, Schneider S, Karras U. Ubbo Visser, Gerald Steinbauer, Alexander Ferrein. Knowledge-based instrumentation and control for competitive industry-inspired robotic domains. In: Künstliche Intelligenz. Springer Berlin Heidelberg. 30th ed. 2016. pp. 289-299

[2] Schneider S, Hegger F, Hochgeschwender N, Dwiputra R, Moriarty A, Berghofer J, Kraetzschmar G. Design and development of a benchmarking testbed for the factory of the future. In: IEEE International Conference on Emerging Technologies and Factory Automation (ETFA): Special Session on Mobile Robotics in the Factory of the Future; 8-11 September 2015; Luxembourg. IEEE; 2015

[3] RoCKIn Project. Project Website [Internet]. 2014. Available from: http://rockinrobotchallenge.eu/ [Accessed: 26 May 2017]

[4] Ahmad A, Awaad I, Amigoni F, Berghofer J, Bischoff R, Bonarini A, Dwiputra R, Fontana G, Hegger F, Hochgeschwender N, Iocchi L, Kraetzschmar G, Lima PU, Matteucci M, Nardi D, Schiaffionati V, Schneider S. RoCKIn Project D1.2 "General Evaluation Criteria, Modules and Metrics for Benchmarking through Competitions". 2014. Available from: http://rockinrobotchallenge.eu/rockin_d1.2.pdf [Accessed: 26 May 2017]

[5] Amigoni F, Bonarini A, Fontana G, Matteucci M, Schaffionati V. To what extent are competitions experiments? A critical view. In: Workshop on Epistemological Issues in Robotics Research and Research Result Evaluation; Hong Kong. ICRA 2014. 05 June 2014. http://rockinrobotchallenge.eu/publications.php

Chapter 5

RoCKIn Benchmarking and Scoring System

Giulio Fontana, Matteo Matteucci,
Francesco Amigoni, Viola Schiaffonati,
Andrea Bonarini and Pedro U. Lima

Additional information is available at the end of the chapter

http://dx.doi.org/10.5772/intechopen.70013

Abstract

The main innovation brought forth by the European Project RoCKIn is the definition, implementation and application to an actual robot competition of the novel paradigm of *benchmarking through competitions*. By doing so, RoCKIn set in motion an evolutionary process to transform robot competitions from successful showcases with limited scientific impact into *benchmarking tools* for the consistent and objective evaluation of the performance of autonomous robot systems. Our work began by revisiting, in the light of the features and limitations of a competition setting, the very foundations of the scientific method; then we built on these by designing a novel type of competitions where the concepts of benchmark and objective performance metrics are the key points; finally, we arrived to the implementation of such concepts in the form of a real-world robot competition. This chapter describes the above process, explaining how each of its several aspects (theoretical, technical, procedural) has been tackled by RoCKIn. Special attention will be devoted to the problems of defining performance metrics and of capturing the *ground truth* needed to reliably assess robot perceptions and actions.

Keywords: benchmarking, robot competition, benchmarking through competitions, performance metrics, ground truth

1. Introduction

The main innovation brought forth by the European Project RoCKIn [1] is the definition, implementation and application, in the form of actual robot competitions, of the novel paradigm of **benchmarking through competitions** [2].

This paradigm is an evolution of the concept of robot competition. The idea is that—alongside the already established roles of demonstration towards the general public and networking event for researchers—competitions for autonomous robots can and should evolve to become benchmarking tools for robot systems. This is why the tests that robots are subjected to, during the RoCKIn competitions, are called *benchmarks*: their aim is in fact to act as reference tasks and activities in which robot performance is evaluated according to well-specified and quantitative metrics. The pioneering work of RoCKIn publicly demonstrated the feasibility of this approach, paving the way to further developments. One of such developments is the *European Robotics League* (*ERL*), an on-going robot competition set up by the European Project RockEU2 [3].

Robot competitions do not easily lend themselves to act as benchmarks, intended as rigorous evaluation procedures to assess the capabilities, reliability, dependability and performance of robot systems in precisely defined settings [4]. First, because the (often frantic) setting of a competition is badly suited to the execution of procedures that require accuracy and care. Second, because concepts that are crucial to experimentation and to benchmarking (such as *repeatability* and *replicability*, defined in Section 2) are difficult to reconcile with the necessary spectacular element of a public competition. As the final objective of RoCKIn was to obtain results that could be transferred into other competitions, such difficulties have been carefully taken into account, and viable solutions have been devised. Both the technical elements and the procedures required by such solutions had to avoid interfering with the execution of the competition.

A first instance of successful infusion of the RoCKIn legacy into other established robot competitions has occurred at the 2016 RoboCup competition held in Leipzig (Germany). In fact, at RoboCup 2016, the aforementioned ERL has been able to both collect benchmarking data from some of the existing RoboCup tests and to incorporate benchmarks directly based on such tests into the European Robotics League.

The first phase of RoCKIn's work consisted of going back to the foundations of the experimental method to carefully reassess the elements that characterize a *scientific experiment*. A further step has been to define the special case of the **benchmarking experiment**, that is, a comparison test which presents some of the features of a scientific experiment. Finally, this analysis formed the foundation for the design and execution of the benchmarks involved in the two challenges of the RoCKIn Competition (RoCKIn@Home and RoCKIn@Work). As explained in other chapters of this book, RoCKIn@Home focuses on a service robot scenario where a robot has to assist a person in her daily life, while RoCKIn@Work is aimed at the shop-floor scenario.

This chapter is dedicated to providing the reader with a summary of the steps composing the path that leads from the scientific foundations (Sections 2 and 3) to the implementation of the RoCKIn Competition, focusing on the methodologies (Sections 4 and 5) and the infrastructure (Section 6) used by RoCKIn to design and execute the Competition. Special attention is to be devoted to the solutions devised by RoCKIn to the problems of defining reliable performance metrics for robot activities, and of capturing the *ground truth* (GT) necessary to apply such metrics in an objective and consistent way.

http://dx.doi.org/10.5772/intechopen.70013

2. Robot competitions as benchmarking tools

Project RoCKIn is based on the idea of **benchmarking through competitions**: that is, of transforming competitions into vessels for experiment-based scientific progress. The successful RoCKIn Competition demonstrated the feasibility of this innovative approach. Besides RoCKIn, the point of view that robotic competitions can (under suitable circumstances and despite some essential differences) be considered as experiments has also emerged elsewhere, both within the academic community and at the level of the European Commission. In particular, competitions are now considered as good vehicles for advancing the state-of-the-art in terms of new algorithms and techniques in the context of a common problem [5–8].

While scientific progress is often related to the concept of experiment, in the majority of cases significant differences exist between experiments and competitions [2]. Just to cite the most obvious, an experiment is aimed at evaluating objectively a specific hypothesis, while a competition is aimed at defining a ranking and winners; for this reason, competitions push towards the development of solutions, while experiments aim at exploring phenomena. Notwithstanding these and other differences, there are a number of reasons for recasting robot competitions as experiments, considering traditional experimental principles (comparison, repeatability, reproducibility, justification, etc.) as guidelines. *Comparison* is to know what has been already done in the field, to avoid the repetition of uninteresting experiments and to get hints on promising issues to tackle. *Reproducibility* is the possibility for independent scientists to verify the results of a given experiment by repeating it, while *repeatability* is the property of an experiment that yields the same outcome when performed at different times and/or in different places. *Justification* and *explanation* deal with the necessity of interpreting experimental data in order to derive correct implications.

Competitions usually provide controlled environments where approaches to solve specific problems can be compared. Furthermore, they require integrated implementations of complete robotic systems, suggesting a new experimental paradigm trying to complement the rigorous evaluation of specific modules in isolation (typical of most laboratory research). RoCKIn set out to prove that an experiment-oriented perspective on competitions can reach the aims of both research and demonstration, while providing a common ground for comparison of different solutions. By reframing robot competitions as experiments via the RoCKIn Competition, the project aimed at increasing their scientific rigour while trying to maintain their distinctive aspects, which are significant and valuable. For instance, competitions are appealing to the participants (people like to compete) and to the general public, in a way that laboratory experiments could never achieve. Competitions are excellent showcases of the current state-of-the-art in research and industry. Competitions push their participants to their creative limits, coordinating to solve difficult problems while doing better than their competitors, ultimately leading to the development and sharing of novel solutions. Competitions promote critical analysis of system performance out of labs. Finally, competitions are a way to share the cost and effort of setting up complex installations among a multitude of participants, making costly experimental setups feasible.

3. Benchmarking experiments

Although competitions can be considered as a way of comparing the performance of robots in a partially controlled environment, their character of being, to some degree, unique events, puts serious limits on the generalizability and replicability of their results. As it has been already noticed [9], robot competitions are not necessarily experimental procedures: on the contrary, some of their features may not fit an assessed experimental methodology. A competition can be considered as a kind of experiment only if its settings and scoring are properly defined.

To define what we intend for experiments, we turn to experimenting practice in computing, which can be intended as the empirical practice to gain and check knowledge about a computing system. It is worth noticing that in this context there are at least five different ways in which the notion of experiment is used [10]. These are ranked below, ordered by increasing complexity of execution and, more importantly, of general scientific significance of the results.

- *Feasibility experiment*. It is the loosest use of the term 'experiment' that can be found in many works reporting and describing new techniques and tools. Typically, the term 'experiment' is used in this case with the meaning of empirical demonstration, intended as the existence of proof of the ability to build a tool or a system.
- *Trial experiment*. This requires the evaluation of various aspects of a system using some predefined variables, which are often measured in laboratory, but can occur also in real contexts of use, possibly given some limitations.
- *Field experiment*. It is similar to trial experiment in its aim of evaluating the performance of a system against some measures, but it takes place outside the laboratory in complex socio-technical contexts of use. The system under investigation is thus tested in a live environment, and features such as performance, usability or robustness are measured.
- *Comparison experiment*. In this case, the term experiment refers to comparing different solutions with the goal of looking for the best solution for a specific problem. Typically, comparison is made in some setup and is based on some measures and criteria to assess the performance. Thus, alternative systems are compared and, to make this comparison as rigorous as possible, standard tests and publicly available data are introduced.
- *Controlled experiment*. It is the golden standard of experimentation of traditional scientific disciplines and refers to the original idea of experiment as controlled experience, where the activity of rigorously controlling (by implementing experimental principles such as reproducibility or repeatability) the factors that are under investigation is central, while eliminating the confounding factors, and allowing for generalization and prediction.

Many existing robot competitions are designed in such a way that their position within the above experimental hierarchy is not higher than field experiments. This cannot be considered as a flaw of such competitions, since they are usually not aimed at being recognized as scientific experiments. On the contrary, the aspiration of the RoCKIn Project has been to define a competition based on tests acting as **benchmarking experiments** [11].

http://dx.doi.org/10.5772/intechopen.70013

In RoCKIn, benchmarking experiments are defined as *a way of performing experimental evaluation, of comparing different systems on a common, predefined, setting and of providing a set of metrics (together with a proper interpretation) to perform an objective evaluation, with the goal of enabling the reproducibility and repeatability of experiments*. The goal of RoCKIn is to devise benchmarking experiments that—according to the rank presented before—can be classified as comparison experiments or even, possibly, controlled experiments.

It is important to point out that the concept of 'objective evaluation' does not rule out human judgement of robot performance, which is often a key tool for performance evaluation (e.g. whenever human-robot interaction is involved). For RoCKIn benchmarks whose performance metrics include evaluation by humans, 'objective evaluation' means the setup of a suitable framework to ensure that human judgement is done according to clearly defined criteria, and that the elements of such judgement are separated and visible (instead of being lumped together in a single score). Several RoCKIn benchmarks (such as the one illustrated in **Figure 1**) involve human-robot interaction.

Additional information about how human judgement is managed in the context of RoCKIn benchmarks is available in Section 5.

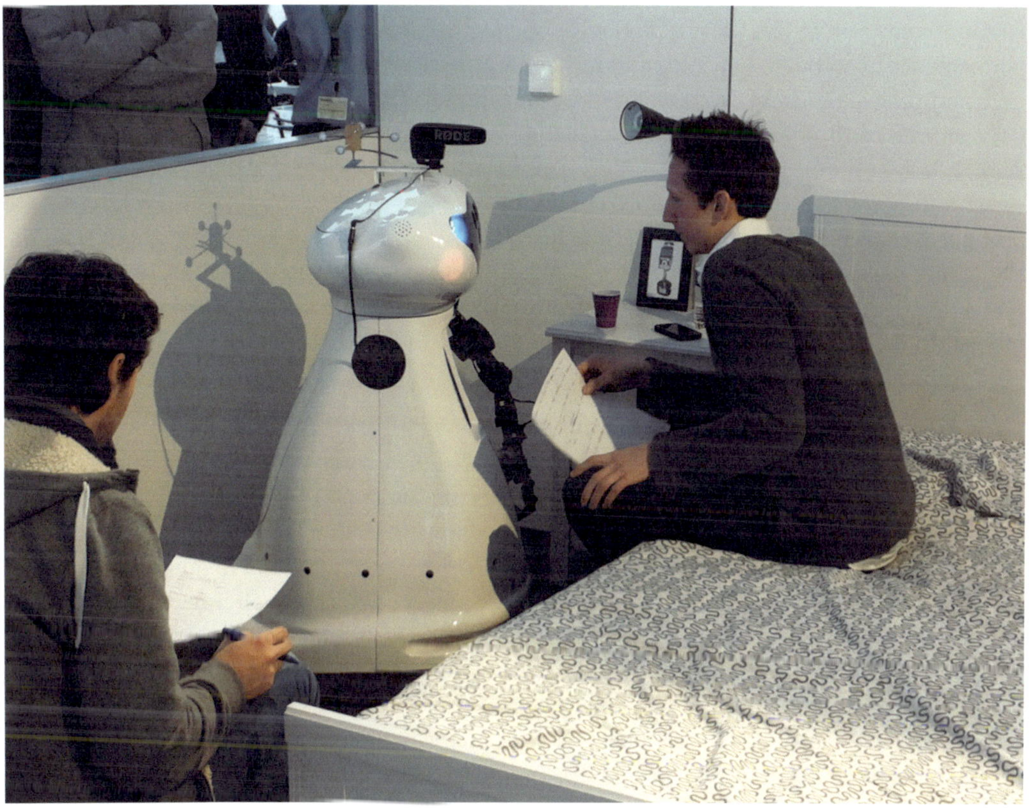

Figure 1. Example of RoCKIn benchmark requiring human robot interaction (at the RoCKIn 2015 Competition in Toulouse).

4. Benchmarking modules and systems

The approach of RoCKIn to benchmarking experiments (in the sense defined in Section 3) is based on the definition of two separate, but interconnected, types of benchmarks [2]:

- **Functionality benchmarks** (**FBM**s), which evaluate the performance of *robot modules* dedicated to specific functionalities, in the context of experiments focused on such functionalities.
- **Task benchmarks** (**TBM**s), which assess the performance of *integrated robot systems* facing complex tasks that require the interaction of different functionalities.

Of the two types, FBMs share more similarities with a scientific experiment. This is due to their stricter control on setting and execution. On the other side, this same feature of functionality benchmarks limits their capability to capture all the important aspects of the overall robot performance in a systemic way.

Focusing on either integrated systems or specific modules is a limit of traditional robot competitions and benchmarks. For instance, RoboCup@Home [12] and RoboCup@Work [13] assess the performance of integrated robot systems executing specific tasks in domestic or factory environments, while the Rawseeds Benchmarking Toolkit [14] is dedicated to assessing the performance of software modules that implement specific functionalities such as self-localization, mapping and SLAM (Simultaneous Localization And Mapping). Unfortunately, focusing only on one of these two approaches (system or module analysis) strongly limits the possibility to gain useful insight about the performance, limitations and shortcomings of a robot system. In particular, evaluating only the performance of integrated systems can identify the best performance for a given application, but it does not provide information about how the single modules are contributing to the global performance, and provides no information about where to spend further development effort in order to improve system performance. On the other side, the good performance of a module in isolation does not necessarily mean that it will perform well when inserted in an integrated system.

The RoCKIn Competition targets both aspects, thus enabling a deeper analysis of a robot system by combining system-level (TBM) and module-level (FBM) benchmarking [15]. Module-level testing has the benefit of focusing only on the specific functionality that a module is devoted to, removing interference due to the performance of other modules which interact with it at the system level. For instance, if the grasping performance of a mobile manipulator is tested by having it to autonomously navigate to the grasping position, visually identify the item to be picked up and finally grasp it, the effectiveness of the grasping functionality is affected by the actual position where the navigation module stopped the robot, and by the precision of the vision module in retrieving the pose and shape of the item. On the other side, if the grasping test is executed by placing the robot in a predefined position and by configuring it with precise information about the item to be picked up, the final result will be almost exclusively due to the performance of the grasping module itself. The first test can be considered as a 'system-level' benchmark, because it involves more than one functionality of the robot; on the contrary, the second test can assess the performance of the grasping module with minimal interference from other modules and a high repeatability, and can thus be classified as 'module-level' benchmark.

http://dx.doi.org/10.5772/intechopen.70013

It must be stressed that there are issues that module-level testing can neither identify nor assess, and nonetheless have a major impact on real-world robot performance. For instance, the interactions among the navigation, vision and grasping modules, which act as disturbance factors in evaluating the performance of the grasping module alone, take a crucial role in defining the actual performance of a robot system in a real setting where grasping is needed. Performing an experiment that excludes such interactions (such as a FBM focused on grasping) implies a major loss of useful information. Here lies the specific worth of system-level robot testing: it is the only way of making system-level properties apparent. We already cited the most obvious of such properties (i.e. direct interactions among modules), but subtler ones exist. One of the most important one is the quality of the integration between modules: experience shows that this is indeed crucial for the capability of a robot to achieve its goal.

Autonomous robots are systems of sufficiently high complexity to make loosely defined emerging properties (such as the aforementioned 'system integration') an important factor in the overall performance of the integrated system. As a consequence, even perfect knowledge of the performance of each robot module does not provide reliable, or sufficient, information to predict the performance of the complete robot once these modules are put together.

The considerations reported in this section can be represented in a matrix form, as shown in **Figure 2**.

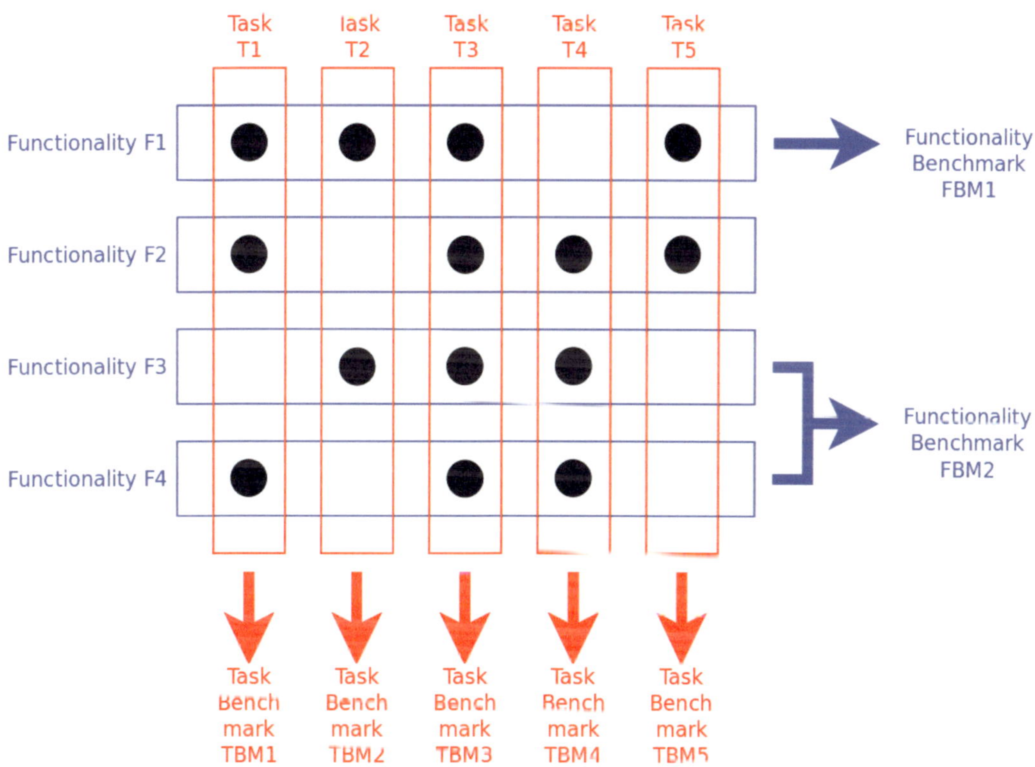

Figure 2. Benchmarking along the horizontal (functionality or module-level) and vertical (task or system-level) directions.

Let us consider an imaginary version of the RoCKIn Competition composed of five tasks (T1, T2, …, T5). **Figure 2** describes such competition as a matrix, showing the tasks as columns while the rows correspond to the functionalities for successfully executing the tasks. For the execution of the whole set of tasks, four different functionalities (F1, F2, …, F4) are required; however, a single task usually requires only a subset of these functionalities. In **Figure 2**, task Tx requires functionality Fy if a black dot is present at the crossing between column x and row y. For instance, task T2 does not require functionalities F2 and F4, while task T4 does not require functionality F1.

Two final observations conclude this section. First of all, while the robot tasks explored by the RoCKIn Competition correspond to the TBMs, the Competition does not necessarily include a FBM for each of the functionalities required by such TBMs. Second, it is conceivable that a functionality benchmark tests more than one functionality at the same time while still allowing to separate their contributions. In **Figure 2**, this happens to FBM2, which tests functionalities F2 and F4.

The reader is invited to compare theory with practice by consulting the descriptions of the specific functionality and task benchmarks composing the RoCKIn Competition, as provided by the Rulebooks of the RoCKIn Competition [16, 17].

5. Performance metrics

Performance metrics are an important element of the task benchmarks (TBMs) and functionality benchmarks (FBMs) presented in Section 4. In particular, their definition has a key role in enabling the benchmarks to act as *benchmarking experiments*, a concept defined in Section 3.

It is not possible to define useful general-purpose benchmark metrics: to be relevant, performance metrics need to be closely related to the specific robot activities under test. For some activities, it is reasonably easy to define suitably objective metrics: this mostly happens when the scope of the activity is limited and very well defined. In the context of RoCKIn, this mainly applies to FBMs. Other times, defining objective metrics is not easy. This happens especially when the activity that the robot is required to perform is complex, composed of different parts and with multiple objectives. For RoCKIn, this typically applies to TBMs.

As already pointed out in Section 3, an especially critical problem is that of defining performance metrics for robot activities that require human robot interaction (for instance, all the task benchmarks of RoCKIn@Home require HRI). In this case, subjective judgement by humans cannot be expunged from performance evaluation; on the contrary, it is crucial to it and must necessarily be part of the metrics. However, this introduces the necessity to establish a framework to guide and interpret subjective judgements, in order to limit their arbitrary elements and to enable their use as elements of consistent performance metrics. Defining such framework has been one of the tasks of RoCKIn; its results will be described in the following of this section. More precisely, Section 5.1 deals with the problem of collecting ground truth data, while Sections 5.2 and 5.3 explain how RoCKIn manages the problem of defining meaningful performance metrics, respectively, for task benchmarks and functionality benchmarks.

5.1. Ground truth

While discussing metrics, a key issue is that of **ground truth** (GT). In fact, in order to define objective performance metrics, it is necessary to get reliable data about the actual activity of the robot. Once available, these data can be compared (depending on the benchmark considered) either with the expected goals, or with robot perceptions. The accuracy of GT data should in any case be sufficiently high that any residual error is much lower than the accuracy required from the robot. In this way, with correctly designed performance metrics, errors in GT data have a negligible effect on the benchmark score.

Some types of GT data are suitable for collection by human referees: for instance, in FBM1 of RoCKIn@Home and RoCKIn@Work (object perception: the robot is required to identify an object presented to it and provide its pose), the actual identity of the object is ascertained by the referee. Other types of data, instead, can only be determined with sufficient precision using special machines: in FBM1, an example of such data is the pose of the object.

For robotics, the pose of an object (either a part of a robot such as the base or the end effector, or an external object such as those used for FBM1) is indeed an especially important type of ground truth. As a consequence, the capability of accurately measuring such data is a key enabler in the development of robot benchmarks. For this reason, RoCKIn collects pose data using a specialized machine [18], the main component of which is a *motion capture* (*mocap*) *system*. While RoCKIn does not specify the type of mocap system but only its performance, the current setup (which will be described in Section 6) is based on a commercial product. The mocap system uses IR-sensing cameras observing the volume of space where the objects to be tracked move, and special reconstruction software fed with the output of the cameras. The system is not capable of localizing objects on their own; instead, it localizes IR-reflecting *markers* affixed to the object. For the mocap system, a set composed of three or more rigidly connected markers can be used to define a *rigid body*, to which a 3-axes reference system is associated. When the system perceives a set of markers corresponding to a known rigid body, it computes and outputs the pose of the rigid body. This output pose, read by software developed for RoCKIn, is the ground truth used to compute the benchmark metrics. In the RoCKIn Competition, the rigid bodies tracked using the motion capture system are associated to **marker sets**, that is, special objects fitted with configurations of markers chosen to maximize tracking accuracy. Examples of RoCKIn marker sets are shown by **Figures 3** and 7. During the execution of RoCKIn benchmarks that require pose measurements, marker sets are affixed to the objects to track and the mocap system used to track the associated rigid bodies. Tracking data from the mocap system are then used for online localization of the object. For instance, FBM2 of RoCKIn@Home is a navigation benchmark where the robot is required to reach a sequence of poses; for this benchmark, then, a marker set is affixed to the base of the robot in order to measure the differences between assigned and actual robot poses.

5.2. TBM metrics: achievements and penalties

The scoring framework for the evaluation of the task benchmarks in the RoCKIn Competition is based on the concept of **performance classes**. The performance class of a robot is determined by the number of **achievements** (or goals) that the robot collects during its execution of

Figure 3. Marker set used for functionality benchmark 'Control' of RoCKIn@Work at the 2015 RoCKIn Competition.

the assigned task. Within each class, ranking is defined according to the number of **penalties** collected by the robots belonging to the class. Penalties are assigned to robots that, in the process of executing the assigned task, make one or more of the errors (which correspond generally to unwanted behaviours) defined by a list which is part of the specifications of the TBM.

More formally, in order to establish the ranking of the robots that execute a specific TBM, the elements of three sets have to be defined. While the contents of these sets are specific to the specific TBM considered, the general semantics is common to all TBMs. The three sets are:

- set **A** = **achievements or goals**, that is, things that the robot *is required to do*: during the execution of the benchmark, an **achievement** is assigned to the robot for each of these;
- set **PB** = **penalizing behaviours**, that is, things that the robot *is required to avoid doing*: during the execution of the benchmark, a **penalty** is assigned to the robot for each of these;
- set **DB** = **disqualifying behaviours**, that is, things that the robot *absolutely must not do*.

The content of each of the sets above must be specified as part of the specifications of the TBM. Then, the ranking of the robots that executed the same TBM is defined according to the following rules:

- The performance class of a robot corresponds to the number of achievements collected by the robot during the execution of the benchmark. Class 0 is the lowest performance class.
- A robot belonging to performance class N is considered as higher in rank than a robot belonging to performance class M whenever $M < N$.
- Among robots belonging to the same performance class, ranking is defined by the number of penalties collected by the robots: if robot R1 has less penalties than robot R2, then R1 is considered as higher in rank than R2.
- Among robots belonging to the same performance class and with the same number of penalties, the ranking of the robot which accomplished the task in a shorter time is considered as higher.

To apply the RoCKIn scoring framework for task benchmarks, the following three-step sorting algorithm is used:

1. if one or more of the disqualifying behaviours of set DB occurred during task execution, the robot gets disqualified, that is, it is assigned at performance class 0 and no further scoring procedures are performed for it;
2. the robot is assigned to performance class X, where X corresponds to the number of goals of set A accomplished by the robot (these sets do not contain repetitions, thus if a given achievement has to be accomplished multiple times, there will be as many distinctive instances of that achievement as required by the task; for instance, if the task requires to serve four guests during dinner, there will be four items in set A, one for each guest); and
3. a penalization is assigned to the robot for each behaviour belonging to set PB that occurred during the execution of the task. Unless clearly specified, it is sufficient that a penalized behaviour occurs once to assign a penalty, and further repetitions of the same behaviour do not lead to additional penalties.

One key property of this scoring system is that a robot that executes a larger part of the task associated to the TBM will always be placed into a higher performance class than a robot that executes a smaller part of the task. The measure of 'what part of the task' a robot accomplished is the subset of set A composed of the achievements assigned to the robot; the metric used to evaluate how large is the 'part of the tasks' accomplished by a robot is the number of elements of such subset, that is, the number of achievements assigned to the robot. Penalties do not change the performance class assigned to a robot and only influence intra-class ranking.

So far, for RoCKIn task benchmarks the assignment of achievements and penalties and the detection of disqualifying behaviours has been performed by human referees. This is an example of how human judgement, if correctly employed, can be part of a ranking procedure without compromising the objectivity of such procedure. In the case of RoCKIn's task benchmarks, the key to such objectivity lies in the precise definition of the elements of the aforementioned sets A, PB and DB, and in the training of the referees. Printed forms prepared with suitable boxes are provided to the referees, in order to guide their work and reduce the probability of mistakes.

It is possible that future benchmarks based on RoCKIn's framework (such as those developed by the on-going European Project RockEU2 [3]), or future implementations of existing benchmarks, will make use of methods different from human judgement to detect goals, penalized behaviours and/or disqualifying behaviours. This will not require any change to the scoring framework described in this section.

As a real-world example of TBM metrics, the remainder of this section describes one of the task benchmarks of the 2015 RoCKIn Competition (Lisbon, Portugal). Interested readers can find a complete description of the benchmark (including much more detail) in the RoCKIn@ Home Rulebook [16].

5.2.1. Example: task benchmark 'Welcoming Visitors'

A person takes the role of *Granny Annie,* a fictional character corresponding to an elderly woman. Granny Annie is helped in her daily activities by her service robot (i.e. the robot under test). In this TBM the robot is required to handle several visitors who arrive at Annie's home and ring the doorbell. The robot has to treat each visitor appropriately, according to the following scenarios:

- *Dr Kimble* is Annie's doctor stopping by to see after her. He is a known acquaintance; the robot lets him in and guides him to the bedroom.
- The *Deli Man* delivers the breakfast; the actual person is changing almost daily, but they all have a Deli Man uniform. The robot guides the Deli Man to the kitchen, and then guides him out again. The robot is supposed to constantly observe the visitor.
- The *Postman* rings the doorbell and delivers mail and a parcel; the actual person is changing almost daily, but they all have a Postman uniform. The robot just receives the deliveries and bids farewell to him.
- An *unknown person,* trying to sell magazine subscriptions, is ringing. The robot will tell him goodbye without letting the person in.

The robot must recognize the visitor by comparing the images from a camera located outside the door to known faces and/or uniforms. Interaction between people and robot is done vocally. Performance evaluation for this TBM is done as follows.

The set A of the achievements is composed by the following elements:

- The robot opens the door when the doorbell is rung by Dr Kimble and correctly identifies him.
- The robot opens the door when the doorbell is rung by the Deli Man and correctly identifies him.
- The robot opens the door when the doorbell is rung by the Postman and correctly identifies him.
- The robot opens the door when the doorbell is rung by an unknown person and correctly identifies the person as such.

http://dx.doi.org/10.5772/intechopen.70013

- The robot exhibits the expected behaviour for interacting with Dr Kimble.
- The robot exhibits the expected behaviour for interacting with the Deli Man.
- The robot exhibits the expected behaviour for interacting with the Postman.
- The robot exhibits the expected behaviour for interacting with an unknown person.

The set PB of *penalized behaviours* is composed by the following elements:

- The robot fails in making the visitor respect the proper rights.
- The robot generates false alarms.
- The robot fails in maintaining the original state of the environment.
- The robot requires extra repetitions of speech.
- The robot bumps into the furniture.
- The robot stops working.

Finally, the set DB of *disqualifying behaviours* is composed by the following elements:

- The robot hits Annie or one of the visitors.
- The robot damages the testbed.

5.3. FBM metrics: benchmark-specific measurements

As explained in Section 2, among the RoCKIn benchmarks, FBMs are those that can be more easily designed to act as *benchmarking experiments* (i.e. *a way of performing experimental evaluation, of comparing different systems on a common, predefined, setting, and of providing a set of metrics—together with a proper interpretation—to perform an objective evaluation, with the goal of enabling the reproducibility and repeatability of experiments*). The reason for this is that FBMs are focused on one (or a very small subset) of the functionalities of a robot, which allows a much more precise definition of the activity that the robot is required to perform with respect to what happens in TBMs.

An important consequence of focusing the benchmarking action towards specific functionalities is that it sometimes enables the benchmark designer to completely eschew evaluation by human referees, thus making the definition of objective metrics easier. As observed in Section 5.2, devising objective performance metrics which include human evaluation is possible, but requires special care. On the other hand, a performance metric based on a well-specified algorithm applied to instrumental measurements of physical quantities is objective by definition. Of course, an objective metric—if badly designed—can nonetheless be a bad indicator of robot performance: however, this is a problem common to any metric.

An example of the 'objective by design' performance metrics described above are those used by RoCKIn's FBMs assessing the physical movements of the robot (or parts of it) through space. These metrics are based on comparisons (according to specific criteria) of the expected motion of the robot (or robot part) with the ground truth pose data produced by the motion capture system introduced in Section 5. Section 6 will show how such system is set up and used in practice.

A consequence of the very specificity of the functionality benchmarks is that it is impossible to define a general scoring framework for FBMs. In fact, the close link between FBMs and a single functionality requires that performance metrics are based on the features of such functionality. For this reason, a general methodology suitable for all FBMs cannot be defined. This differs markedly from what has been done for task benchmarks in Section 5.2, where a common framework for the performance metrics for TBMs, based on the concept of *performance classes*, was presented.

As a real-world example of FBM metrics, the following of this section describes one of the functionality benchmarks used in the 2015 RoCKIn Competition (Lisbon, Portugal). Interested readers can find a complete description of the benchmark (including much more detail) in the RoCKIn@Work Rulebook [17].

5.3.1. Example: functionality benchmark 'Control'

This functionality benchmark assesses the capability of a robot to control the manipulator's (and the mobile platform's) motion in a continuous manner. The robot has to follow a given path in the Cartesian space using the tip of a *marker set*, that is, a special object (shown in **Figure 3**) which can be precisely localized in space using RoCKIn's motion capture system.

More precisely, the mocap system is used to measure the deviation between the assigned path and the path actually followed by the tip of the marker set due to the movements of the robot's end effector. In the 2015 RoCKIn Competition, the given path could be a segment of a straight line or a portion of a sine function.

Without going into the procedural details of the benchmark, we focus here on the accuracy metric used to assess control performance. Let us define:

- $r(l) = (x_r(l), y_r(l))$ the parametric representation of the actual robot path,
- $t(l) = (x_t(l), y_t(l))$ the parametric representation of the given (target) path,

where l is a parameter ranging from 0 to 1. Then, the accuracy metric is

$$\frac{1}{N}\sum_{l \in L_{sampled}} d(r(l), t(l)) \tag{1}$$

where $L_{sampled}$ is a subset of the set L_{gt} of values of l in correspondence to which is available a location measurement from the ground truth system, $N = |L_{sampled}|$ and $d()$ represents the Euclidean distance between two points.

As anticipated, for this FBM the process of collecting the necessary data and computing the accuracy metric is entirely performed by machines; no human intervention is required.

5.4. RoCKIn benchmarking system

The RoCKIn benchmarking system is the infrastructure supporting the activities of the RoCKIn Competition that are directly related to benchmarking. The setup described in this section corresponds to the one used by at the 2015 RoCKIn Competition held in Lisbon, Portugal.

The system is composed of two interconnected but separate subsystems: one dedicated to the RoCKIn@Home benchmarks, and the other to the RoCKIn@Work benchmarks. This is due to the fact that in Lisbon the benchmarks of these two challenges were running in parallel due to the time constraints of the competition. In a less demanding setting, it would be possible to lower the number of components of the RoCKIn benchmarking system by relaxing the constraint of being capable of managing one RoCKIn@Home benchmark and one RoCKIn@Work benchmark at the same time.

6. System architecture

The architecture of the RoCKin benchmarking system is shown in **Figure 4**.

As shown in **Figure 4**, for each of the two challenges (RoCKIn@Home and RoCKIn@Work) the system includes three main computers. These are:

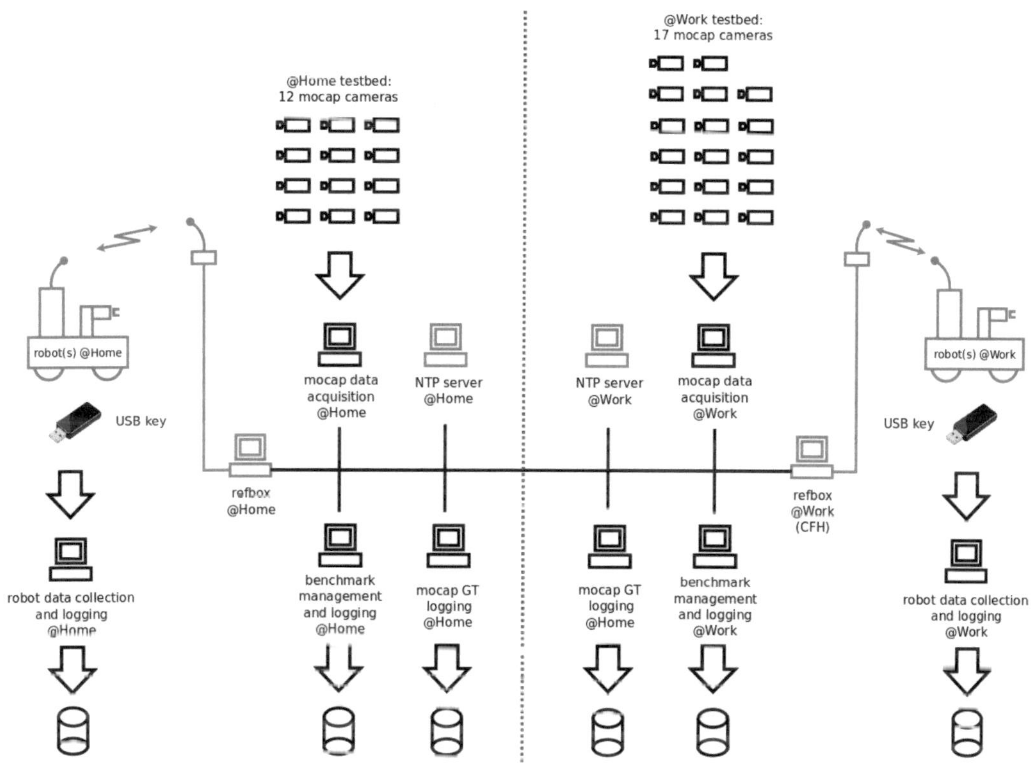

Figure 4. Architecture of the RoCKIn benchmarking system used at the 2015 RoCKIn Competition held in Lisbon, Portugal.

1. one computer acquiring motion capture data from the special cameras of the mocap system and streaming it to the other machines;
2. one computer processing the mocap data streamed by the former to extract and log ground truth (GT) pose data; and
3. one computer managing the benchmarks (which also logs the data related to their execution).

One additional computer is used to collect and save robot-generated data, logged by the robots on USB keys during the execution of the benchmark. These USB keys are physically brought by the teams to the referees immediately after each benchmark.

In the end, the number of machines involved and the complexity of their interconnections is fairly high. This is due to several factors, including the fact that the software of the mocap system requires the Windows operating system, while all other machines are Linux-based, and the fact that the various subsystems have been developed (and physically brought to the Competition) by different partners of project RoCKIn. **Figure 5** shows the 'benchmarking table' hosting part of the PCs used for benchmarking for RoCKIn@Home at the 2015 RoCKIn Competition. A similar working area was used for RoCKIn@Work.

Figure 5. PCs used for RoCKIn@Home benchmarks at the 2015 RoCKIn Competition.

To operate, the RoCKIn benchmarking system also needs to interact with external systems, shown in grey in **Figure 4**. Interactions occur over TCP/IP networks, which include wireless segments. The systems external to the RoCKIn benchmarking system shown (in grey) in **Figure 4** are:

1. the robot under test;
2. the Referee Box (also called Central Factory Hub or CFH in RoCKIn@Work), which organizes competition activities, interfaces with devices belonging to the testbed and interacts with human referees; and
3. the NTP (Network Time Protocol) server with which all the PCs in **Figure 4** (including those on board of the robot) have to synchronize.

Synchronization is important for the correct execution of the RoCKIn benchmarks, for two reasons. First, because benchmark execution requires to associate and compare data generated by different sources, which can only be done if such data is correctly time-stamped. Secondly, because RoCKIn records datasets comprising both robot-generated data (e.g. sensor streams) and data generated externally to the robot (e.g. ground truth): for the datasets to be usable, all such data streams must therefore share the same time base.

A consequence of the synchronization constraints described above is that, in order to execute one of RoCKIn's benchmarks, a robot must precisely align its own internal clock to the clock of the RoCKIn NTP server. The Referee Box checks for misalignments and only starts the benchmark when these have been reduced below a predefined threshold. To help participating teams to perform such adjustment automatically, RoCKIn recommended installation on the robots of a software package called *Chrony* and provided a suitable configuration file for it.

6.1. Motion capture setup

To be able to benchmark actual robot performance, RoCKIn needs to collect ground truth data. For RoCKIn an especially important category of GT data is that describing the pose of objects and robots in space. As anticipated in Section 5.1, RoCKIn captures such data using a custom hardware and software system based on a commercial motion capture (mocap) system. The mocap system used at the 2015 Competition is called OptiTrack and is manufactured by Natural Point. OptiTrack relies on special infrared 'smart' cameras and proprietary software (running partly on the cameras and partly on a PC) to detect the location of IR-reflective markers. **Figure 6** shows an example of how RoCKIn used such cameras for its activities.

A set of at least three markers having fixed distance between each other can be defined as a *rigid body* in the OptiTrack mocap system. The system can then track the 6DOF pose of all defined rigid bodies and stream the data, which are subsequently collected by special software running on one of the machines in **Figure 4** (*mocap GT logging*). RoCKIn benchmarks make use of this to track the pose of special objects called *marker sets*. When the marker set is rigidly affixed to a robot component, it is possible to reconstruct the pose of the component from tracking data.

Figure 6. Part of the motion capture cameras used to cover the RoCKIn@Home area at the RoCKIn Camp held in Peccioli (Italy) in 2015.

For instance, to track robots, RoCKIn uses marker sets composed of a planar base (made of 4-mm-thick plywood) fitted with five spherical markers, shown in **Figure 7**.

The locations of the markers of the marker set maximize the distances between markers while keeping the marker set reasonably compact. Most importantly, such locations have been carefully chosen to ensure that inter-marker distances are all significantly different. This is necessary to prevent ambiguity, which may cause severe fluctuations in reconstructed pose. The marker set of **Figure 7** is used, for instance, during the execution of the functional benchmark 'Navigation' of RoCKIn@Home. This FBM (already presented in Section 5.1) requires that the robot navigates through the environment to reach, in order, a series of waypoints specified in terms of position and heading.

Practical experience at RoCKIn events (Camps and Competitions) showed that, unfortunately, obtaining a good setup of the motion capture system requires significant experience. Especially critical are the choice of camera locations and the tuning of the system for optimum performance, also keeping in mind the effect of local lighting. These aspects become less and less critical as the number of cameras increase; however, given the considerable cost of each camera, RoCKIn tried to keep their number as low as possible (though, as shown in **Figure 4**, still not very low in absolute terms; this high number is a direct consequence of the large observed

Figure 7. Marker set used to track robots at the 2015 RoCKIn Competition. To get an idea of the its dimensions, the reader can consider that each of the five spherical markers has a diameter of 19 mm, and that the base of the marker set fits within a circle with a diameter of 170 mm.

volumes). In the end, RoCKIn always operated close to the edge of the performance envelope of the OptiTrack mocap system, thus minimizing the cost of the system but paying a price in terms of difficulty of setup and expertise required for successful installation and parameter setting.

Another difficulty of using a motion capture system in a temporary setting (such as a robot competition) is that it is difficult to ensure the required consistency over time of relative camera locations. Even small changes in these locations can, in fact, greatly affect the performance of the mocap system. For this reason, for instance, mobile installations such as the tripod-based one shown in **Figure 6** are not acceptable for competitions. The solution chosen by RoCKIn makes use of an overhead truss, which of course is much heavier, larger and more difficult to mount and dismantle. **Figure 8** shows a rendering of the truss mounted above and around the RoCKIn@Home testbed at the 2015 competition; a similar truss was used for the RoCKIn@Work area.

6.2. Motion capture usage

At the 2015 RoCKIn Competition, each participating team was required to mount a marker set of the type shown in **Figure 7** on the top of their robot (fitting the marker sets on top minimizes

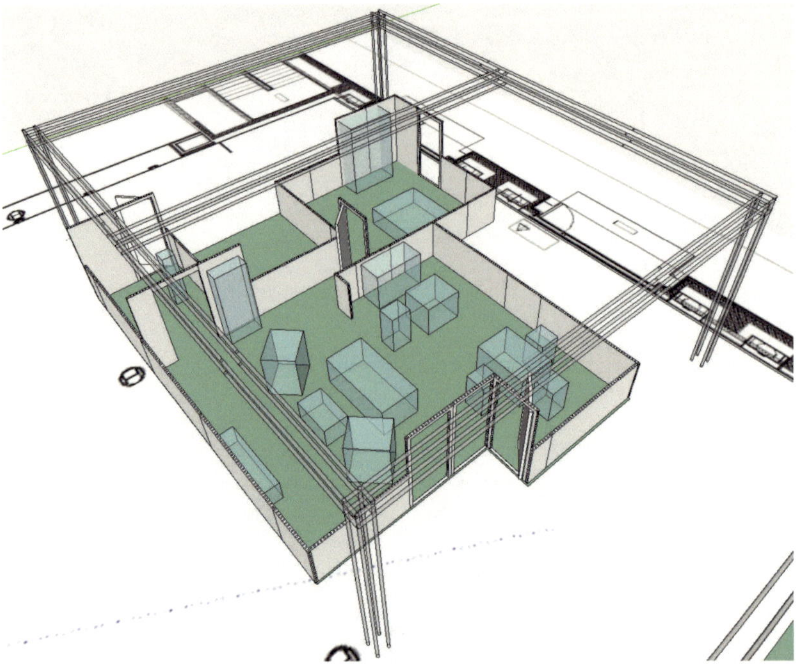

Figure 8. Rendering of the overhead truss used to support motion capture cameras around and above the RoCKIn@ Home testbed of the 2015 Competition.

occlusions due to robot parts), in a roughly horizontal orientation (thus minimizing occlusions between markers, as explained later), with the arrow-shaped marker base pointing forward according to the robot's own odometry reference frame (the shape of the marker set has been chosen to make pointing obvious). To facilitate mounting, marker sets are provided with holes, and CAD models of the marker set base are available.

RoCKIn acquired the transform between the odometry reference frame of each participating robot and the reference frame of the marker set mounted on it. Such transforms have been used, during the execution of the benchmarks, to reconstruct robot pose (according to the robot's own coordinate system) from motion capture data. In this way, ground truth data produced were directly comparable with odometry data logged by the robot, thus facilitating the assessment of the robot's odometric performance. The procedure to acquire the transforms required each team, in turn, to place their robots on the ground in a predefined location, with the X, Y axes of the robot's odometry frame aligned in a predefined way.

Beyond the ones mounted on robots, additional marker sets are used as parts of the setup for specific benchmarks. One of these special purpose marker sets, used for the 'Control' FBM of RoCKIn@Work, is shown in **Figure 3**. Other specialized marker sets are used for the 'Object perception' FBMs of RoCKIn@Home and RoCKIn@Work. This benchmark, already presented in Section 5.1, requires that the robot identifies and localizes a series of objects placed in front of it. **Figure 9** illustrates the elements of the experimental setup for FBM1, while **Figure 10** shows their use during the execution of the benchmark.

http://dx.doi.org/10.5772/intechopen.70013

Figure 9. Setup for functional benchmark 1 (Object Perception) at the 2014 RoCKIn Competition in Toulouse, France. The (red) motion capture cameras used to track the objects presented to the robot are mounted on the metal truss adjacent to the table.

Figure 10. Example of execution of functional benchmark 1 (RoCKIn@Work version) using the setup of **Figure 9**.

In the setup of **Figure 9**, both the table top where the objects are placed for perception and a small wooden tablet supporting the objects (visible in **Figure 10**) are actually marker sets. This way, the mocap system can be used to find the transform between the reference systems associated to them; by combining such transform with the (previously recorded) transform between the object's own reference systems and the tablet's, it is possible to obtain the pose of the object with reference to the table top, to be compared to the reconstructed pose provided by the robot. The AR (augmented reality) markers visible on the table top in **Figure 10** are used to define the 2D reference system that the robot is required to use for reconstructed poses.

It is interesting to point out that the marker set of **Figure 7**, used at the 2015 Competition, is planar and thus significantly simpler to build than the '3D' version used at the 2014 Competition (which was similar to the marker set of **Figure 3**). This change is deliberate, and comes from practical experience. Its goal is to minimize occlusions between markers: in the difficult lighting conditions of the 2014 Competition (a white, partially light-transparent tent, in the open), such occlusions compromised localization performance, requiring 'on the fly' modification of the marker sets. Thus, for the 2015 Competition we designed new marker sets taking better advantage of the known features of the relative positions of mocap cameras and markers. In fact, cameras are significantly higher from the ground than markers, thanks to the mounting points on the overhead truss: therefore, with a marker set with all the markers are on the same horizontal plane, critical occlusions only tend to occur when the markers are perceived by mocap cameras that are already too far from the marker set to provide useful localization data.

7. Conclusions

The differences between scientific experiments and robot competitions are many and significant. Project RoCKIn set out to a difficult task: that of developing methodologies to design novel robot competitions whose tests, without losing the traditional role of technology showcases, could at the same time act as veritable *benchmarking experiments*.

During the life of the project, the above methodologies have been developed; a competition based on them—the *RoCKIn Competition*—has been designed; and two editions of it have been successfully held (in 2014 and 2015). This means that RoCKIn has reached its goal. Most importantly, it means that the way to further, fruitful developments in the field of robot benchmarking is open. Some of these developments are already on-going.

During the course of this chapter, the whole process leading to this result has been retraced; encompassing—without burdening the reader with excessive detail—the range from theoretical foundations to real world implementation. For what concerns the latter, particular attention has been devoted to the key problem of collecting ground truth data.

Author details

Giulio Fontana[1], Matteo Matteucci[1*], Francesco Amigoni[1], Viola Schiaffonati[1], Andrea Bonarini[1] and Pedro U. Lima[2]

*Address all correspondence to: matteo.matteucci@polimi.it

1 Dipartimento di Elettronica, Informazione e Bioingegneria, Politecnico di Milano, Italy

2 Institute for Systems and Robotics, Instituto Superior Técnico, U. Lisboa, Portugal

References

[1] RoCKIn Project. Project Website [Internet]. 2014. Available from: http://rockinrobotchallenge.eu/ [Accessed: 26 May 2017]

[2] Ahmad A, Awaad I, Amigoni F, Berghofer J, Bischoff R, Bonarini A, Dwiputra R, Fontana G, Hegger F, Hochgeschwender N, Iocchi L, Kraetzschmar G, Lima PU, Matteucci M, Nardi D, Schiaffionati V, Schneider S. RoCKIn Project D1.2 "General Evaluation Criteria, Modules and Metrics for Benchmarking through Competitions" [Internet]. 2014. Available from: http://rockinrobotchallenge.eu/rockin_d1.2.pdf [Accessed: 26 May 2017]

[3] RockEU2 Project – European Robotics League. Project Website [Internet]. 2016. Available from: http://sparc-robotics.eu/the-european-robotics-league/ [Accessed: 26 May 2017]

[4] Amigoni F, Schiaffonati V. Models and experiments in robotics. In: Magnani L, Bortolotti T, editors. Handbook of Model-Based Sciences. Springer; 2017. pp. 799-915

[5] Anderson M, Jenkins O, Osentoski S. Recasting robotics challenges as experiments. IEEE Robotics Automation Magazine. 2011;**18**(2):10-11

[6] Cohn AG, Dechter R, Lakemeyer G. The competition section: A new paper category. Artificial Intelligence. 2011;**175**:iii

[7] Holz D, Iocchi L, van der Zant T. Benchmarking intelligent service robots through scientific competitions: The RoboCup@Home approach. In: Proceeding AAAI Spring Symposium on Designing Intelligent Robots: Reintegrating AI II;Palo Alto, USA. 2013. pp. 27-32

[8] Smart B. Competitions, challenges, or journal papers. IEEE Robotics Automation Magazine. 2012;**19**(1):14

[9] Croce D, Castellucci G, Bastianelli E. Structured learning for semantic role labeling. Intelligenza Artificiale. 2012;**6**(2):163-176

[10] Tedre M, Moisseinen N. Experiments in computing: A survey. The Scientific World Journal. 2014;2014:11 p. Article ID 549398. DOI: 10.1155/2014/549398

[11] Amigoni F, Bonarini A, Fontana G, Matteucci M, Schiaffonati V. To what extent are competitions experiments? A critical view. In: Proceedings of the ICRA 2014 Workshop on Epistemological Issues in Robotics Research and Research Result Evaluation; Hong Kong. 2014

[12] RoboCup@Home Competition [Internet]. 2017. Available from: http://www.robocup.org/domains/3 [Accessed: 26 May 2017]

[13] RoboCup@Work Competition [Internet]. 2017. Available from: http://www.robocup.org/leagues/16 [Accessed: 26 May 2017]

[14] RAWSEEDS Project. Project Website [Internet]. 2004. Available from: http://www.rawseeds.org/ [Accessed: 26 May 2017]

[15] Amigoni F, Bastianelli E, Berghofer J, Bonarini A, Fontana G, Hochgeschwender N, Iocchi L, Kraetzschmar G, Lima P, Matteucci M, Miraldo P, Nardi D, Schiaffonati V. Competitions for benchmarking: Task and functionality scoring. IEEE Robotics & Automation Magazine. 2015;**22**(3):53-61

[16] RoCKIn Project. RoCKIn@Home Rulebook [Internet]. 2015. Available from: http://rockinrobotchallenge.eu/rockin_d2.1.3.pdf [Accessed: 26 May 2017]

[17] RoCKIn Project. RoCKIn@Work Rulebook [Internet]. 2015. Available from: http://rockinrobotchallenge.eu/rockin_d2.1.6.pdf [Accessed: 26 May 2017]

[18] RoCKIn Project. "Description of Ground Truth System V2", Deliverable D2.1.8 [Internet]. 2015. Available from: http://rockinrobotchallenge.eu/rockin_d2.1.8.pdf [Accessed: 26 May 2017]

Chapter 6

RoCKIn: Impact on Future Markets for Robotics

Rainer Bischoff and Tim Friedrich

Additional information is available at the end of the chapter

http://dx.doi.org/10.5772/intechopen.70012

Abstract

The goal of the project "Robot Competitions Kick Innovation in Cognitive Systems and Robotics" (RoCKIn), funded by the European Commission under its 7th Framework Program, has been to speed up the progress toward smarter robots through scientific competitions. Two challenges have been selected for the competitions due to their high relevance and impact on Europe's societal and industrial needs: domestic service robots ("RoCKIn@Home") and innovative robot applications in industry ("RoCKIn@Work"). The history and reasoning behind the chosen task and functionality benchmarks in RoCKIn are explained by providing an insight from the *International Federation of Robotics* and an analysis on RoCKIn's impact on the industrial robot market domain is carried out. To paint a broad picture, RoCKIn is compared to other robot competitions and similarities, differences and challenges those competitions share are pointed out. Some industrial robot market requirements and the way RoCKIn addressed them are explained. Strength and weaknesses of the project in regard to their market impact are emphasized and it is shown how these were continued and addressed by RoCKIn's successor European Robotics League (ERL).

Keywords: robotics, robot competitions, benchmarking, domestic robots, industrial robots

1. Introduction

This chapter gives a brief overview on the current situation of the robot market. It discusses the potential impact robot competitions could have on this market by not only focusing on RoCKIn's (Robot Competitions Kick Innovation in Cognitive Systems and Robotics) contributions, but also on its successor, the European Robotics League (ERL), and robot competitions and benchmarking in general. It is divided into two main sections:

- Competitions in industrial market domains
- Analyses on market impact

This chapter focuses mainly on industrial robot competitions and their impact on the industrial robot market, though some of the information provided is also applicable for service robot competitions and their impact on the service robot market.

The following sections briefly discuss the current situation of the robotics market, the general requirements of end users and the way RoCKIn addressed them. It is further shown how the industrial and service robot markets are potentially influenced by robot competitions and some of their long-term benefits. Some success stories beginning in robot competitions are introduced before concluding with an outlook on upcoming robot competitions and benchmarking efforts throughout Europe.

2. Competitions in industrial market domains

According to the *International Federation of Robotics* (*IFR*), the robot market can be divided into the two major areas: industrial and service robotics. Classification within those areas follows and extends the *Standard Industrial Classification of All Economic Activities (ISIC) revision 4*. For industrial robotics alone over a dozen different fields are analyzed. The most interesting in the context of RoCKIn@Work are *Manufacturing*, *Electrical/electronics*, and *Automotive*. From an application perspective, *Handling operations / Machine tending*, and *Assembling and disassembling* inspired the RoCKIn@Work benchmarks. In RoCKIn@Home, the service robot, application areas *Robots for domestic tasks* and *Elderly and handicap assistance* were represented through the benchmarks.

With an increasing need for robot-based automation, solutions come from the demand to objectively compare different solutions. As part of a solution development or purchase process, end users usually compare the data sheets of different robot makers/vendors and sometimes even create their own benchmarks to be able to find the most promising solution. In particular, the automotive industry creates benchmarks that are known to the robot makers, but not to the general public. They compare and judge the robot performance in secrecy and share the results with the robot makers. Typical benchmarking criteria are measurable quantities, such as cycle time, path accuracy, pose repeatability, mean-time between failures, cost, etc. Soft factors such as perceived product quality, brand image, long-term customer relationships, etc. are also playing a role here.

The big question is how to objectively compare different solutions. There are a lot of different robot challenges, benchmarks, and competitions set in industrial market domains. Often they intend to address one specific problem. Only very few of them try to solve different tasks of varying complexity. Most of them have in common that a winner will be determined by one or more referees, but they lack an objective scoring and comparability of different functionalities. This is a key difference to the idea of RoCKIn@Work, where the single functionality of a robotic system, as well as the overall performance on a task level are scored through objective metrics that rely on data gathered throughout tests.

This section briefly describes a few industrial end-user problems and outlines the importance and impact of robot competitions have had so far and how they might even get a stronger influence.

2.1. Robot market challenges

As outlined in the introduction, the worldwide robot market is growing fast and companies that intend to utilize robots in their factories to automate their production processes, especially SMEs that have not yet started to automate their processes, are confronted with an increasing amount of possibilities to do so. The number of robot companies is growing rapidly and they often provide very specialized solutions to common problems. It is very hard for companies to select a solution that fits best their needs because there is no objective way to compare the solutions offered in the market. This is where industrial robot competitions start to become attractive for industry and where they set themselves apart from other well established competition formats. They no longer focus only on the playful character of a competition, but also intend to provide solutions for a problem occurring in the real world. This helps SMEs, with only little experience in robotics so far, to get an idea of the state of the art in robotics and about possible solutions for a production challenge they face in their own factories.

2.2. Different industrial robot competitions

This section outlines some of the robot competitions set in the industrial domain with a high impact in the robotics community.

They have in common that the quality of the participating teams is very high. It is apparent that the competitions are not taken lightly and a lot of effort is spent on their preparation. One more indication for this is that a lot of Ph.D students participate in them and make them part of their daily work. They are often accompanied by younger undergraduate students, thus sharing knowledge and providing them with hands-on experience outside the usual university curriculum.

What is missing in most of them is that they do not provide a setup for mobile manipulation, something which is becoming of increasing importance in *Factory of the Future* scenarios [1]. Instead they focus on a very complex setup to challenge participants, either with a lot of different objects to manipulate, or with a very complex task as subset of a real world application.

2.2.1. Airbus Shopfloor Challenge

This challenge was held for the first time at ICRA 2016 in Stockholm. Teams had to solve several rounds of a simplified drilling task on an artefact representing part of the aircraft fuselage. Success was measured based on the number of holes drilled within a specified time and accuracy. Apart from a cash prize for the winning team, much more important was the possibility for the winners to develop their idea for commercial application within Airbus. Further details are provided in Ref. [2].

2.2.2. *Amazon Robotics Challenge*

The goal of the Amazon Robotics Challenge [3] is to strengthen the ties between the industrial and academic communities and promote shared and open solutions to some of the big problems in unstructured automation of so far logistics/handling use cases. The challenge consists of three different tasks. In the *Pick* task, teams have to remove target items from storage and place them into boxes. During the *Stow* task, teams have to pick target items from totes and place them into storage. This is followed by a final round, where all items are first stowed and then selected items are picked into boxes.

2.2.3. *Bayer Robotics Competition*

The year 2016 marked the start of another robotics competition that is organized and funded through Grants4Tech, an initiative of Bayer AG, aiming at improving production processes in the life science industry [4]. Very similar to the Amazon Robotics Challenge, the competition aims at strengthening the bond between industry and academic robotic research communities and enthusiasts. The task at hand is derived from a real world problem. At Bayer, everyday sampling of powder for quality control from a drum of incoming raw materials is done hundreds of time. To find a solution how this process could be automated, the robotics competition was founded. The task is broken down into subtasks, some of which are a complex task in itself. At the time of writing, the competition is ongoing, therefore no conclusion can be drawn yet.

2.2.4. *European Robotics Challenge (EuRoC)*

The European robotics challenge is a project funded through the European Union's Seventh Framework Programme. It aims at sharpening the focus of European manufacturing through a number of application experiments, while adopting an innovative approach that ensures comparative performance evaluation. It consists of three industry-relevant challenges:

- Reconfigurable interactive manufacturing cell
- Shop floor logistics, and manipulation
- Plant servicing and inspection

Out of the before mentioned competitions, EuRoC is the only one that, like RoCKIn, utilizes benchmarking as performance measurement. Another key difference is that this competition involves mobile manipulation and runs in different stages over a period of four years, whereas the other competitions are yearly events. More information can be found at [5].

2.3. RoCKIn@Work

The RoCKIn competition, as described in detail in previous chapters, sets itself apart from the other industrial robot competitions described above. Its main focus is not only the competition part, but also benchmarking. The project is, to the best of our knowledge, the first robot competition that allows performance assessment of a robot functionalities and abilities relying on objective metrics through comparison of recorded data with a ground-truth system. The idea is to use competitions as a tool for scientific experiments to foster scientific progress and innovation which has been very well perceived in the research and industrial community.

Although the direct impact of RoCKIn to industry is not measurable, it is important to have a competition that engages students to have a closer look at the hardware and software they develop or use. Building a modular system where single functionalities can be tested without interference of other parts of the robotic system is still, as far as robot competitions go, unique to RoCKIn. RoCKIn therefore was an ideal entry format to competitions and to benchmarking. Younger students, maybe in their second or third year at university, had a chance to join a team and be able to contribute, either through a software/hardware module of their own, or through modifying existing functionality.

Those skills learned and the hands-on experience is very valuable on a job application. Personal contact to a representative of industry, in RoCKIn's case, the KUKA Roboter GmbH, opened up possibilities for participants to apply for internships or to write their final thesis in one of the major robotics companies worldwide because hands-on experience significantly increases the success rate of applicants. It can be observed that students with a background in robotics competitions, either RoCKIn or others, are in general more resilient and committed to their assigned task. Their work, be it their final assignment from university or work done during an internship, is often of higher quality than the work of those who do not have this experience. During their time in the company, they need less support from their supporting employee. Often they are able to work better on their own and in teams.

2.4. European Robotics League: Industrial Robots

The European Robotics League (ERL) is the successor to RoCKIn. It is part of the larger project RockEU2, funded through the Horizon 2020 framework program of the European Commission. The *ERL Industrial Robots* (*ERL-IR*) league is successor to RoCKIn@Work. As explained in the first chapter of this book, the main difference between the ERL and RoCKIn is distinction between *major* and *local* tournaments.

This concept was introduced to tackle problems that were grounded in the competition format itself. In RoCKIn, and also other robot competitions, often fast solutions, or hacking some new functionality, was common during competitions and consequently led to misbehavior of the system. The tight schedules during competitions simply did not allow for any major changes in hardware or software. By introducing *local tournaments* teams should be able to spend enough time preparing for the benchmarks to avoid any major errors in their system. They should also get the time to fix minor problems between benchmark execution the "right" way, avoiding hacking wherever possible. In the end, this should lead to much more robust systems, able to face the challenges set by the benchmarks repeatedly.

At the moment the ERL addresses rather undergraduate students than PhDs. For this reason, the *local tournaments* provide another advantage over usual competitions: training and direct interaction between experts and participants. Compared to the *ERL Service Robots* (*ERL-SR*) league, ERL-IR has even younger participants, not accompanied by many PhD students. This might be the case because in ERL-SR, there is no standardized platform unlike in ERL-IR where until now all teams participated with a KUKA youBot, making it the de-facto standard platform for the league. The advantage of local tournaments in ERL-IR now presents itself through the possibility of direct sharing of software or knowledge between participants because they use the same hardware platform. Even helping out with parts, tools or reducing time spend on searching for software errors because the error is already known by someone else which is common among the teams. This exchange is very fruitful and helps fostering progress toward smarter robots through knowledge sharing and community building.

3. Analyses on market impact

3.1. Industrial requirements

Industrial applications have a set of requirements that are common among most of them. These requirements include:

- Setup time
- Cycle time
- Dependability
- Technology readiness
- Cost

RoCKIn addressed those topics through the design of the RoCKIn@Work testbed and the definition of the task and functionality benchmarks. To make sure that *setup time* is kept to a minimum the time allowed for each team to prepare their robot for a task was fixed by providing them with a time slot before each benchmark execution. *Cycle time* was addressed by imposing a time limit on the benchmark run (typically 10 min). The time for a single step in a task Benchmark, for example recognizing an object, has not been fixed, but was thought of as possibility by the consortium. It has not been implemented because the performance of a single functionality was assessed during the functional benchmarks.

In the case of *dependability*, the teams had to command their robots to execute every benchmark multiple times at several pre-fixed starting times. This ensured that teams could spend time between the benchmarking runs on improving the robot setup and fix some smaller errors, leading to a more dependable system at the end of the competition. Although programming ("hacking") during competitions can be seen negatively, it at least teaches the team members valuable lessons about keeping deadlines, performance requirements, and quality issues.

The RoCKIn project had only little resources compared to other competitions. The teams participating in the competition also had only few resources. Consequently, the solutions found to tackle the challenges had to be *low-cost*. For example, hardware changes most often came through 3D-printed additions to the main robot platform used. This is a very cost-effective way to change a hardware setup compared to milled metal parts which are still standard in industry.

3.2. Analyses of current situation

The strengths of RoCKIn were manifold. One that stood out was the engagement teams showed toward the scientific and benchmarking aspects of the competition. Working in a common problem domain and sharing knowledge toward the goal of solving the task, more than once prevailed over "winning" the competition. To experience and test their robot system in a well-defined manner outside their own lab and in competition with others was seen as very valuable, both for participants as well as for the project.

What was missing before RoCKIn and has not yet been solved to a satisfying degree is specification of benchmarking challenges. The benchmarks defined by RoCKIn are a step in the right direction, but do not yet reach the impact that would be necessary to engage more partners from industry. Being very complicated up to a point where most teams were not able to execute a complete benchmark is noteworthy. Expectations by the consortium were maybe too high in the beginning, but the rules were also not flexible enough to take the participants skill level into account for a specific benchmark. One more thing still lacking is a process of tech transfer to industry. In RoCKIn, as well as in other robot competitions, there is still no clearly defined process as how software modules could be transferred to a real world application, something that industry would be very interested in. Further the possibilities to evaluate performance of single modules or functionalities are still limited. The infrastructure to roll out benchmarking in a broader context does not yet exist.

To approach this problem, RoCKIn laid a very good foundation on which the ERL is now building on. Having a clear specification of a testbed for benchmarking is utilized by the ERL to replicate them in laboratories all over Europe. A certification process ensures that the new testbed is able to execute specific task or functionality benchmarks. Rule evolution during the competition years is going to address different participant skill levels and will allow for more possible combinations of functionality into task benchmarks. The scientific part of the competition is going to become considerably more prominent and the amount of benchmarking data useful for other researcher will increase.

The growing number of robot competitions is both blessing and curse. Too many competitions specializing on a particular problem might lead to more specialized solutions instead of generalized solutions. Specialized solutions are something companies are already very proficient in. To advance the state of the art, those solutions need to become applicable to a wider range of applications to keep up with market expectations. Production has to be much more flexible in the future, which is why applications should not be tailored to one specific use case anymore.

4. Success stories

This section shows two success stories that had their beginning in robot competitions.

The first is the Amazon Robotics Challenge, which is explained in Section 2. It started out as the Amazon Picking Challenge as part of the ICRA Robot Challenges in 2015. This was preceded by Amazon acquiring Kiva Systems, which itself became Amazon Robotics. Kiva

Systems was a company realizing efficient warehouse logistics, founded by one of the participants of the RoboCup small-size league.

The second success story is the one of the *RoboCup Standard Platform League*. The league started out using the Sony AIBO as standard platform. After discontinuation of AIBO, a new company, Aldebaran Robotics, was founded by one of the league's participants. They developed the small humanoid robot NAO which replaced AIBO as standard platform. After acquisition of Aldebaran Robotics by SoftBank, it was announced that in 2017 their robot Pepper will become the new standard platform for the newly introduced *RoboCup@Home Standard Platform League*.

These stories show that robot competitions are worth much more than one would expect on first sight. It is the team spirit and the ever growing community that pushes talented people to pursue something extraordinary which might not be possible without it.

5. Conclusion and outlook

Throughout Europe, robot competitions gain momentum and start to attract more researchers every year. It is expected that new competitions, which focus more closely on end-user needs and which provide solutions with a higher Technology Readiness Level (TRL) than current ones, will appear. The possibility to transfer developed technologies to a market ready application more easily will raise interest of more companies. The increasing complexity will lead to more (semi-)professional teams competing in competitions. It is very likely that more teams will be supported by companies through collaboration or financial backup.

The professionalization of robot competitions, as can be seen by other big players like Amazon, Airbus, or Bayer creating their own competitions, is going to result in more success stories like the ones pointed out in this chapter.

Most of the current industrial robot competitions share the thought of an open community and highly flexible and exchangeable software and hardware. At the point where they find together and start to truly interact with each other, maybe even through a co-organized robot competition, it can be expected that a lot of innovative products will become available on the market. This will further increase the competitiveness of Europe as one of the major players in the Fourth Industrial Revolution [6].

Author details

Rainer Bischoff* and Tim Friedrich

*Address all correspondence to: rainer.bischoff@kuka.com

KUKA Roboter GmbH, Germany

References

[1] European Commission. Factories of the Future [Internet]. Available from: https://ec.europa.eu/research/industrial_technologies/factories-of-the-future_en.html [Accessed: February 19, 2017]

[2] Airbus. Airbus Shopfloor Challenge [Internet]. Available from: http://www.airbus-group.com/int/en/people-careers/Working-for-Airbus-Group/Airbus-Shopfloor-Challenge-2016.html [Accessed: February 19, 2016]

[3] Amazon. Amazon Robotics Challenge [Internet]. Available from: https://www.amazonrobotics.com/#/pickingchallenge [Accessed: February 19, 2016]

[4] Grants4Tech. Bayer Robotics Challenge [Internet]. Available from: https://grants4tech.bayer.com/home/ [Accessed: February 19, 2016]

[5] EuRoC. European Robotics Challenges [Internet]. Available from: http://www.euroc-project.eu/index.php?id=euroc_project [Accessed: February 19, 2017]

[6] Bundesministerium für Bildung und Forschung. Industrie 4.0 [Internet]. Available from: http:// http://www.plattform-i40.de/I40/Navigation/DE/Home/home.html [Accessed: February 19, 2017]